C000165712

Water Survival Guide

Mastering the Art of Finding, Collecting, Treating, Storing, and Thriving Off the Grid with Water for Emergencies and Outdoor Adventures

Table of Contents

Introduction

Water is the giver of life from which everything emerges. Therefore, it is one of the first resources you need when going off-grid or in an outdoor survival emergency. Gathering, treating, and storing your water are all skills you need to master for a truly self-sustained existence, free from the shackles of societal systems that aim to hold you back. As a survivalist, prepper, outdoors person, or an off-grid homesteader, you can benefit from the mountain of enlightening information you are about to explore.

When you've finished this book, you'll have an in-depth education on the numerous ways you can gather and clean water. Moreover, you will have focused your survival eye needed to transform any space into a hub to thrive. By understanding natural processes and different biomes, you can tap into the flow of the elements to manifest an abundance of this life-giving and sustaining liquid. Once you internalize some basic principles and techniques, your water survival skills will be fine-tuned and elite. The mix of theoretical knowledge and practical advice gives you a complete view of the intricacies of working with water and the many considerations people forget when they run their bath water.

With a new-found respect for water and its uses, you will unlock ways to think about sustainability. This text explores hydro geography and how the land can be used to your advantage when seeking this essential natural resource. You will also learn how to purify water and make sure that you are kept healthy and strong. Furthermore, you will obtain the knowledge to collect rain and dew in a multitude of different ways. You will even learn how to work with water in the tundra by debunking some of the myths of snow and ice melting while learning the best ways to gain water in these freezing environments.

With a focus on conservation, you will be shown how to gather water in a permanent homestead or when you are on the move. You will be equipped with the skill to survive in settings ranging from water-scarce to water-abundant by learning how to transport and store it for long-term use. Prepare to unfold the multifaceted reality of a resource often overlooked and undervalued as you walk the path of water survival outdoors and thrive off the grid.

Chapter 1: Water: The Essence of Life

While exploring the intricacies of life, you encounter a pivotal element threading through the essence of existence – water. It's a valuable resource embodying the life force that sustains and nurtures all living beings. Not only that, but the Earth's ecosystem also is majorly regulated by water, creating striking and serene landscapes and areas where living beings can inhabit and thrive.

From conception onward, water orchestrates the symphony of human life. Although the human body is 60% water, several vital organs like the heart, brain, and lungs comprise 75 to 80% water. Look around, and you'll find water as the fundamental building block on which life flourishes. It's interlaced with the fabric of human civilization. Ancient cultures revered seas, oceans, and lakes as sacred entities as water promoted sustenance and expansion. To the present day, you can find scriptures and accounts of historians from several civilizations depicting The Nile, Ganges, Tigris, and Euphrates as cradles for great civilizations, shaping the course of history and facilitating the growth of human societies.

Water is the elixir of life.
https://pixabay.com/photos/water-droplets-dewdrops-water-955929/

Water's subtle yet profound role in daily life is mostly overlooked in this technologically driven era where humans are carrying out space explorations on an exponential level and driving industries through cutting-edge technologies. It quenches thirst, keeps the body hydrated, helps detoxify, irrigates fields, generates power, and fuels industries. Despite its omnipresence, the precious nature of water often escapes everyone's notice until its scarcity looms on the horizon.

Water itself is the elixir of life, as most living beings depend on this precious source to stay alive. It isn't just something everyone uses. It's the critical ingredient that keeps life ticking on Earth. From tiny bugs to big animals, water is required for everyone. It's like the builder that makes homes for all the different plants and animals.

Right from the very start of your life, water has been there with you. It's like a cozy bed in a mommy's tummy, helping you grow. Water helps people stay healthy and strong. It's not just about the body. Water shapes how people live together, making communities and traditions possible. Think about the rivers and oceans – they're like big lifelines for human beings. Human ancestors knew this, too. They set up homes close to rivers because they knew rivers were like nature's supermarkets, providing food and water. Even now, in the modern world, water is a critical resource. It helps grow food, keeps living beings hydrated, and powers most industrial processes that shape this era.

Life's Essential Drink

Water is the universal drink for all living things. Plants, animals, and humans need water to survive. It's the basic stuff that keeps your body working. Without water, body functions will deteriorate and ultimately cease to exist.

Keeping Things Cool

Water is nature's air conditioner when the weather gets too hot. It helps cool down the Earth and keeps living things from overheating. It's essential to maintain the ecosystem through natural processes like rainfall.

Growing and Thriving

For plants, water is their food. It helps them grow, make their own food through photosynthesis, and stay green and healthy. Likewise, every other living being needs water to thrive, contributing to a food web from which all animal species benefit.

Quenching Thirst

No matter what beverage, juice, or sugary drink you consume, water is the only source that quenches your thirst. Although its chemical composition is simple, the intestines can easily absorb it, reaching every body part to become body fluids like blood and facilitate metabolic processes.

Essential for Hygiene

Water is not just for drinking. It's also the best liquid for washing. Humans use it to take baths, keep homes tidy, and keep the dishes sparkling. Replacing water as a cleaning substance is nearly impossible. Without it, maintaining hygiene and cleanliness will become nearly impossible.

Powering the World

Beyond keeping life thriving on this green planet, water is used worldwide to power hydroelectric dams. Although solar and wind electricity generation is making waves in terms of clean energy, hydroelectric power generation is still the primary source of electricity that powers homes and runs the gadgets you love to use.

As water is necessary for life itself, you must use it carefully. Pollution and wasting water can harm the environment and make it challenging for everyone, including humans, to have enough clean water. Massive urbanization, the use of clean water for industrial purposes, and the addition of pollutants into streams and rivers are contributing to water waste.

Water is not just a drink. It's a lifeline for all living things. For humans, it goes beyond survival, and it's part of what makes your lives healthy, comfortable, and meaningful. Taking care of water is caring for yourselves and the planet you call home.

The Human Body and Water

Cell Function and Structure

Water is the building material for your cells. It helps form the structure of your cells and ensures they function correctly. It allows the cells to have the right shape, down to the molecular level. Water maintains the cell turgidity and makes body fluids that maintain cell structure and function.

Digestive Dynamo

Why do most people say you should drink water before meals? It's because water is a crucial player in your digestion. It breaks down your food, making it easier for your body to absorb all the good stuff and eliminate the stuff you don't need. Water is absorbed easily through the intestines and is essential to maintain the body's water weight.

Temperature Tamer

You start to sweat when you feel a rise in body temperature after playing sports or on a sunny day. It is your body's way of cooling down, and water is the main ingredient in sweat. So, in a way, water is your built-in air conditioner.

Water is the main ingredient in sweat.
Unsplash https://unsplash.com/photos/persons-eyes-looking-on-left-side-mE6e5-5jLu8

Joint Lubrication

Have you ever thought about how your knees or elbows move so smoothly? Thank water for that. It lubricates your joints, making it easier for them to bend and flex without any squeaky sounds. After absorption in the intestines, water goes through a series of processes, forming the synovial fluid, which acts as a cushion and a shock absorber. It reduces friction in the joints and surrounding cartilage while moving, keeping the bones from rubbing together.

Kidney's Cleaning Crew

Your kidneys are your body's janitor, cleaning up the waste and extra stuff you don't need. Water is their trusty sidekick, helping flush out all the things your body doesn't want to hang around. Most of these unwanted substances enter the body through food consumption, and some chemicals are released as a by-product of processes occurring within the body. All these toxins are flushed through the kidneys in the form of urine.

Blood Flow Booster

Your blood is mostly water, and water helps the blood flow smoothly through your veins and arteries. Good blood flow means all body parts get the necessary oxygen and nutrients. Without adequate water levels, the vessels fail to maintain a consistent diameter, affecting blood flow and reducing the flow of nutrients throughout the body. This increases the chance of blood clots, which can trigger other chronic conditions.

Muscle Function

Water is the fuel that powers your muscles. When you're active, your muscles use water to function correctly. So, whether you're playing soccer or doing cartwheels, water is your secret energy source. Besides enabling free movement, water also promotes muscle gain as it acts as a carrier, delivering materials required for essential proteins and glycogen synthesis. The body's main energy sources are proteins, glycogen, and related materials.

Skin Savior

Ever notice how your skin looks better when you drink enough water? That's because water keeps your skin hydrated, making it look healthy and glowing. Staying hydrated locks in the moisture, which aids in maintaining skin elasticity. This increased elasticity prevents the skin from sagging, reduces wrinkles, and prevents fine lines from developing.

Lastly, water also supports the production of collagen, a substance produced by the body that promotes the development of new skin cells by replacing dead skin cells. It's the best beauty treatment, and it's free.

Brain Buffer

Even if your brain needs water, it helps you think clearly, concentrate, and stay alert. Water facilitates neural connectivity, allowing the brain to communicate effectively. It also clears out the toxin build-up within the brain, optimizing its function. So, if you're ever feeling a bit foggy, a glass of water might be just what your brain needs.

Water is your body's MVP (Most Valuable Player) of bodily functions. It's not just about quenching your thirst; it's the unsung hero behind the scenes, ensuring everything in your body runs smoothly and efficiently. In the realm of essential needs, its scarcity casts a looming shadow on human health. The critical importance of water becomes starkly evident when you consider how brief your survival would be without it.

Life without Water

Threat to Survival

Humans can survive without food for a relatively extended period, but the same cannot be said for water. The body's dependence on water is immediate and critical, with survival jeopardized after just a few days without this precious resource.

Dehydration's Stealthy Onset

The onset of dehydration, a consequence of insufficient water intake, is swift. The human body loses water through daily activities, including breathing, sweating, and waste elimination. These losses escalate in a water scarcity situation, leading to a rapid decline in bodily functions. Over time, these functions will halt, creating a life-threatening situation.

Impact on Physical Health

As discussed earlier, water scarcity contributes to a cascade of health issues. Dehydration impairs the body's ability to regulate temperature, increasing the risk of heat-related illnesses. It also affects the circulatory system, reducing blood volume and increasing potential cardiovascular complications. Furthermore, body fluids like cerebrospinal fluid (essential for the brain) and lymph (required for immune system functioning) won't be produced in the required amounts, compromising cognitive function and weakening the immune defenses. As the body's metabolism slows down and the immune defenses weaken, it opens a window for chronic illnesses, infectious diseases, and medical conditions to affect the body.

Water-Related Diseases

The scarcity of clean water often forces communities to rely on contaminated sources. This, in turn, leads to waterborne diseases, posing severe health threats to populations facing water shortages. Common waterborne diseases include cholera, dysentery, giardiasis, typhoid, scabies, amebiasis, hepatitis, and parasitic infections if the water source is contaminated.

Mental and Cognitive Impacts

Dehydration extends beyond physical health, impacting cognitive function. Even mild dehydration can result in decreased alertness, difficulty concentrating, and impaired short-term memory. Prolonged water scarcity exacerbates these effects, hindering overall mental well-being.

Vulnerability of Specific Groups

Infants, the elderly, and individuals with pre-existing health conditions are particularly vulnerable to the health impacts of water scarcity. Their physiological fragility amplifies the risks associated with dehydration and waterborne diseases.

Aggravation of Existing Health Disparities

Water scarcity exacerbates existing health disparities, disproportionately affecting communities with limited access to clean water.

Socioeconomic Ramifications

Beyond the immediate health concerns, water scarcity amplifies socioeconomic challenges, such as increased healthcare costs, reduced productivity, and a strain on communities. In the face of these issues, the detrimental effects of water scarcity on human health are not just hypothetical scenarios but pressing realities. Recognizing the urgency of ensuring equitable access to clean water becomes paramount to safeguarding human health and well-being.

Food and Water in Life

Survival Imperatives

In situations where water is limited, it immediately impacts the body's fundamental physiological processes, like hydration and nourishment, which are crucial for sustaining life. The absence or scarcity of water poses an imminent threat, making it crucial to find and secure access to water for survival. This water scarcity affects humans and the entire food chain, from microorganisms and plants to animals. Virtually every other species on this planet has a critical role in maintaining this food chain. For example, birds and flying insects are crucial for pollination, and certain species of bacteria and fungi are responsible for decomposing dead organic matter, giving back to the Earth, and so on. When water scarcity peaks, there's a disruption in this complex chain of events, resulting in survival issues for every living being within the ecosystem.

Physiological Vulnerability

As you already know, the human body is an intricate system with tightly interlinked physiological processes. Deprivation of one essential resource, whether for hydration or nourishment, rapidly disrupts these processes. For example, lack of hydration affects temperature regulation, circulation, and immune function. When one of these life-fueling processes within the body doesn't function adequately, it affects every other body function. Timely provision of water and food is essential to keep the body up and running in optimum condition.

Disease Susceptibility

The scarcity of clean resources heightens the risk of disease. Contaminated or insufficient resources can spread waterborne illnesses, malnutrition, and other health issues. This susceptibility to disease threatens public health, necessitating urgent measures to secure access to clean and sufficient resources.

Vulnerability of Specific Groups

Infants, the elderly, or individuals with pre-existing health conditions are particularly vulnerable to the immediate health impacts of resource scarcity. The unique physiological needs of the body and the processes involved get disrupted without the provision of water. If water and food scarcity persist, it can further deteriorate the metabolic processes and even lead to life-threatening situations.

Likewise, underprivileged communities in hot and arid climates struggle to access water and food. This inadequate access causes health to deteriorate rapidly, further burdening healthcare costs. Inadequate provision or scarcity of water and food takes a massive toll on communities like this as they are already struggling to make ends meet.

Impact on Habitat

Likewise, water scarcity also affects other sources of food. For example, if streams in a region dry up and the groundwater levels plummet, the habitat changes drastically, affecting all living beings. From microorganisms to plants, wildlife, and humans, the area will slowly become inhabitable, eliminating virtually every form of life.

Environmental Adaptation Challenges

Both human and natural systems face challenges in adapting to abrupt changes in resource availability. Ecosystems may struggle to cope with shifts in water availability or changes in agricultural practices. Human systems also need to change quickly when resource supply changes.

This detailed exploration highlights the urgency of securing access to critical resources for survival, the vulnerability of physiological processes, the increased risk of diseases, the unique susceptibility of specific populations, the societal and economic repercussions, and the challenges associated with environmental adaptation. Recognizing and addressing these aspects is crucial for effective resource management and the well-being of individuals and communities.

As you reach the culmination of our exploration, the profound significance of water as synonymous with life becomes undeniably clear. The journey through the intricate web of water's roles – from sustaining physiological processes to fostering societal well-being – reveals why this precious resource is often reverently termed 'life-giving.'

Water isn't just a resource; it's the universal elixir that breathes life into everything. Every living creature depends on water for survival and vitality, from the tiniest creatures to the tallest trees.

Throughout the chapter, you've read how water is more than just a drink. It's the foundation of human life, from the earliest days of development to the sustenance of communities and the building blocks of civilizations. It's the nourishing force that shapes the very course of your life. The essence of water as life is not just a metaphor but a fundamental truth. Its deprivation brings forth consequences that resonate from your individual experience to the societal level. As a steward of this precious resource, you must cherish and safeguard water, not just as an environmental duty but as a commitment to the very essence of your life. Through understanding, appreciation, and collective action, you can preserve the vitality of water for generations to come.

In the coming chapters, you'll uncover everything related to water, the tools required to harvest water naturally and serve as a comprehensive water survival guide to follow that can keep you going in unexpected situations and the harshest conditions.

Chapter 2: Hydro-Geography

If you've ever traveled alone in remote areas, you've likely had to find water by the second or third day, at the latest. Finding water isn't usually hard unless you're in an arid area. Typically, you just need to check the quality and treat visible water from big rivers or lakes. The key to finding water in nature is paying attention. When you're in an emergency, the initial reaction is often panic and stress. However, getting hysterical makes things worse, so it's crucial that you stay calm. Being prepared helps you to keep a cool head.

Don't be too worried if you don't see obvious water sources like rivers, lakes, streams, or puddles nearby. Take a few deep breaths. Look out for the signs that nature provides. Since every living thing needs water, the presence of any form of life indicates that there may be usable water close by.

Any form of life indicates that there may be usable water close by.
https://pixabay.com/photos/raindrops-sheets-ladybug-574971/

Understanding Terrain Features for Finding Water

In arid regions, readily available drinking water is rare. Potential water sources can sometimes only be noticed when physically stepping into them. Additionally, crucial signs such as soil layers, indicator plants, and animals can often be identified only in close proximity to a water source, making observation important.

To find promising areas, it's essential to consider the geological characteristics of the environment. Water flows downhill and collects or goes underground at the lowest point. Surveying the surroundings, you can easily identify signs of existing surface water or subterranean water levels.

When examining geological features, it's helpful to know that rock, confining soil layers, and similar structures continue underground in the same pattern as they appear above ground. Disregard the soil layers above a confining layer and visualize how it extends below them.

For instance, if you estimate an intersection of two rocky mountainsides to be beneath some erosion rubble between the slopes, you can mark this point as a potential runoff point. Remember that water may not necessarily break through to the surface but may flow off in the sediment many feet below. Despite this, these natural reservoirs have a significant advantage as they are not affected by evaporation due to the lack of sunlight and can hold water for decades after the last rains.

In certain situations, water accumulates above the water table when it can't drain away quickly. Generally, the water level of open bodies indicates the local groundwater level, lying between a confining soil layer (aquiclude) and surface water level. Trenches in a mountainside suggest the course water can take, indicating the location of water runoff either above or below ground. Canyons or gorges signify water erosion following heavy precipitation events. Follow the dry riverbed until you come across a barrier; if there's a subterranean water reservoir, it might be there. A mountain saddle without natural runoff may contain an open mountain lake or water reservoir under the sediment, with very slow runoff through the bedrock.

Overlapping cliffs with deep valleys represent potential points for a surface river or pond. In a depression in an extensive plain, if there's any subterranean water, it's likely to be found at the lowest point of the depression.

Detecting Water Presence: Animal Tracks and Indicator Plants

Tracking wild animals is often a reliable method for finding water in a natural environment. However, some mammals, especially those in arid regions, have adapted to alternative water sources or can endure long periods without water, making animal tracks less dependable. An extreme example is the desert-dwelling kangaroo rat, which recycles condensed water from its breath in its nasal mucous membrane. Larger mammals may travel great distances to reach the nearest watering hole.

Stories circulate about insects and their relationship to water, but these should be approached with caution. While bees and mosquitoes are known to depend on water, this is mainly true for species in moderate and humid climates. Wind can blow individual animals or swarms into water-free zones where they can survive for weeks or months without water.

Indicator Plants and Animals

- **Ferns, Mosses, and Horsetails:** These plants depend on water for their development, and their widespread global presence makes them valuable indicators of moist or intermittently liquid soil. When you come across areas abundant with ferns, mosses, and horsetails, it signals the likelihood of nearby water sources.

- **Fast-Growing Water Meadow Trees (e.g., poplar, willow):** The characteristics of fast-growing water meadow trees, such as poplar and willow, reveal important information about the surrounding environment. Their shallow roots and preference for temperate zones make them reliable indicators, suggesting the presence of moist soil or surface water in the vicinity.

- **Pandanus:** Found on beaches and prevalent in Oceania's rivers, lakes, marshlands, and freshwater lenses, the presence of pandanus plants is a significant clue. These plants serve as indicators, particularly on beaches, hinting at the potential existence of freshwater lenses beneath the surface.

- **Taro and other (large) Araceae:** The large foliage of taro and similar plants is a clear indicator of moisture in the soil. Commonly found in Southeast Asia near rivers, lakes, and marshlands, their presence signals areas with consistently moist conditions, providing valuable information for water seekers.

- **Reeds, Bulrushes:** Thriving in eutrophic waters globally, reeds and bulrushes are reliable indicators of permanently moist soil. Their presence signifies areas with a high likelihood of liquid water, offering essential clues for those searching for water sources.
- **Tree Ferns:** Surviving only short dry spells, tree ferns found in tropical and subtropical regions indicate the presence of consistently moist soil. When you encounter these ferns, it suggests a local environment with a stable water supply.
- **Rats:** With a global distribution, rats are often found near slow-flowing rivers, lakes, and canals. While not directly indicating the presence of water, their proximity to water bodies can be a relevant consideration, especially when evaluating the overall ecosystem for survival needs.
- **(Water) Snails:** The larger species of snails, equipped with an epiphragm, are often found in moist soil with a high water table worldwide. Their presence indicates areas with accessible water sources, as these snails thrive in environments where water is readily available.
- **Cockatoos, Pigeons, Macaws, etc.:** Mass gatherings of birds like cockatoos, pigeons, and macaws are audible and visible signs of water from long distances. These birds are commonly found near rivers, lakes, and marshlands globally, making their presence a reliable indicator of nearby water sources.

In the Australian outback, bees are not a reliable indicator for water proximity. Many insects and arachnids can survive with rare, microscopically small dew drops for their development stages and water supply.

Certain creatures are particularly useful water indicators, either in liquid form or contained in the soil. These include stationary organisms like plants and highly mobile animals that congregate in large numbers, such as birds.

Indicators for Moist Soil

- In temperate regions, reeds and rush pads are vital water indicators growing in habitats with very moist soils. Their presence usually suggests the availability of liquid water nearby.

- Globally, reeds or cattails are easily identifiable by their long, broad leaves and indicate wet to marshy soils. They can also serve as a water substitute.

- Around the equator in Asia and Africa, taro and other large-leaved Araceae are significant water indicators, thriving in wet conditions.

- Tropical tree ferns, unable to endure long dry spells, settle in the permanently moist ground on hillsides and valleys, indicating subterranean water flows in mountainous regions.

- Even desert-dwelling palm trees, needing water due to fluctuating water tables, can be found in areas without surface water. Their presence suggests the likelihood of water, even in seemingly barren places.

Identifying Potential Groundwater Sources

1. Springs

The most straightforward way to access water is through springs, where subterranean water naturally emerges above ground. Springs can take various forms, such as "rheocrene," where water trickles or gushes out. Often, springs flow into ponds or marshes when the groundwater table is above ground level. Seepage springs are smaller bodies of water without a clear inlet or outlet lying on confining soil layers. Recognizing likely locations involves understanding the geological structures of the landscape.

Indicators of Reliable Springs

A reliable spring may have a relatively low water temperature (around 41 to 50°F [5 to 10°C] in temperate zones, approximately 68°F [20°C] in the tropics) compared to the ambient temperature. Torrents may carry more water beneath the surface than visible, especially in areas with loose soil. A confining soil layer beneath loose stones can store and discharge water in lake- or river-like structures without being apparent above ground.

2. Wells

Wells are artificial attempts to reach the groundwater table. While typical wells may not need detailed instructions, it's crucial to remember that wells are often private property, requiring permission from the owner before using the water. The term "well" encompasses any artificial structure reaching the groundwater table, including simple holes dug to a depth of one to two meters (three to seven feet).

Wells are built to reach groundwater.

PePeEfe, CC BY-SA 4.0 <https://creativecommons.org/licenses/by-sa/4.0>, via Wikimedia Commons: https://commons.wikimedia.org/wiki/File:Water_well_-_Coste%C8%99ti_-_Romania.jpg

Infiltration Wells

Digging may be worthwhile in regions where geological features or other reliable factors suggest the water table is close to the surface. In subtropical regions, water might not be extractable as a liquid but instead collected as moist earth or mud from the unsaturated zone using a suction well. A high groundwater table is indicated by the presence of nearby bodies of water, even if polluted. When digging, pay attention to foul smells from the soil layers, as they could mean the water is polluted.

Important Considerations

- Wells on private property require permission before use.
- Infiltration wells make sense in areas where geological features or reliable indicators suggest a close water table.
- The proximity of a body of water helps estimate the groundwater level.

- Foul smells during digging may indicate polluted water, requiring adjustments to the well location.

Understanding these aspects can improve your ability to locate and access groundwater in a range of terrains and conditions.

Suction Wells

When the sediment lacks enough moisture for liquid water extraction, constructing a suction well is worth considering. Cover the pit's bottom with plant material or clothing fabric, placing a cane, plastic pipe, or similar object inside. Refill the pit with tightly packed soil, allowing liquid water to be sucked into the material and, subsequently, through the cane or tube into your mouth. If this method proves unsuccessful, swallowing moist earth directly or squeezing it through fabric may be necessary. In extreme cases, you may have to distill the water.

Freshwater Lenses

While the term "freshwater lens" may evoke images of underwater photography, it actually denotes a vital phenomenon that safeguards drinking water on islands and secluded beaches. To grasp the dynamics of this freshwater source, you need to delve into the actual concept of density. In essence, dissolving a soluble substance in water leads to the breakdown of molecules or charged ions, dispersing within the water molecules and altering their density. Picture making an impromptu water filter from a combination of coarse gravel and fine sand. The addition of sand doubles the weight but barely affects the volume, showcasing the impact on density.

Swap the sand for table salt, and you'll have a different outcome. The salt dissolves, splitting into Na^+ ions and Cl^- ions (dissociation), increasing density. This disparity in density explains why saltwater outweighs freshwater. A liter of freshwater weighs around 1 kilogram, while seawater of the same volume is about 30 grams heavier. Similar to how Styrofoam floats while lead sinks, freshwater (lower density) rests on top of saltwater. This phenomenon, known as a freshwater lens or Ghyben-Herzberg lens, forms in porous soils like sand, especially where they meet the sea, given occasional rainfall.

Unlike conventional groundwater, freshwater lenses don't rely on impermeable layers; the saltwater beneath the island acts as a confining layer. Notably, freshwater lens levels are higher than sea level. When seeking fresh water on the beach, avoid lower sections and explore the island plateau's highest points for pools or water indicators.

True freshwater lenses may be absent on smaller islands, replaced by brackish water. Salinity assessment becomes crucial in such cases. Meanwhile, submarine freshwater sources on mainland beaches, driven by subterranean rivers, are discernible through geological structures like mountainous trenches or depressions sloping toward the sea.

Collecting water from these sources resembles freshwater lens methods, but it's closer to the beach due to constant interior flux. Contrary to myths, sand doesn't "filter" salt, and opaque streaks in seawater indicate submarine springs. Inland freshwater sources emptying into the sea are prerequisites for beach freshwater. If inland water isn't present, anything dug near the sea is likely to be brackish or salty.

Movement of Water Across Landscapes

Water movement across landscapes is a dynamic process influenced by various factors. Understanding these patterns will make finding water sources in different terrains much easier.

Surface Water Flow

Surface water flow refers to water movement across the Earth's surface. This can occur through various channels, including rivers, streams, and runoff from precipitation. The direction and speed of the surface water flow are influenced by the land's topography. Water follows the path of least resistance, often carving out valleys and channels as it flows downhill.

Recognizing surface water flow patterns involves observing the land's contours and identifying potential watercourses. Understanding the interconnectedness of surface water systems aids in predicting where water is likely to accumulate or flow during rainfall.

Underground Water Flow

Underground water movement, also known as subsurface flow, occurs beneath the Earth's surface within aquifers and permeable rock formations. This subsurface water flow is critical for sustaining groundwater levels and surface water features. Identifying underground water movement involves assessing geological formations and the characteristics of aquifers.

Springs, where groundwater emerges at the surface, are indicators of subsurface water flow. Recognizing the relationship between surface and underground water movement provides valuable insights into overall water availability in a given area.

How Landscape Features Impact Water Movement

Slopes and Gradients

The slope and gradient of the land significantly influence the direction and speed of water movement. Water naturally flows downhill, following the contours of the terrain. Understanding slope dynamics helps predict where water may accumulate or form channels.

Steep slopes contribute to rapid surface water flow, potentially leading to erosion and the formation of valleys. Conversely, gentle slopes may result in slower water movement, allowing for water retention and soil moisture.

Soil Permeability

Soil permeability, or the ability of soil to transmit water, is a crucial factor affecting water movement. Different soil types exhibit varying permeabilities, influencing how water is absorbed, retained, or drained. Sandy soils, for instance, typically have high permeability, allowing water to pass through quickly.

Recognizing soil permeability helps in assessing the likelihood of water retention in the soil and the potential for groundwater recharge. It also aids in understanding how vegetation interacts with soil moisture, providing additional clues for water detection.

Common Misconceptions and Myths

Finding water can be tricky, and plenty of myths may lead you down the wrong path. This section is about setting the record straight on some common misunderstandings and ensuring you know the right way to find water in different places.

Debunking Myths for Different Terrains

- **Deserts**: You may have heard that digging in low spots in the desert will reveal water. Well, it's not that simple. Water doesn't always collect in low areas. Understanding the local geography, like wadis (dry riverbeds) or underground water sources, is more reliable.

- **Forests:** It's a mistake to assume that all parts of a forest have easy access to water just because it's a lush environment. Reliable water indicators, like animal tracks leading to water, are more trustworthy than general assumptions.

Evidence-Based Approaches

- **Smart Techniques:** Forget about mystical methods or using divining rods. Trusting scientific, evidence-based techniques is far more reliable. Knowing your terrain, understanding hydrological principles, and recognizing reliable visual clues are much more reliable.

- **Visual Clues:** Instead of relying on rumors, it's important to be able to identify visual clues in the landscape. Knowing how to spot potential indicators, like specific vegetation patterns, geological formations, and animal behavior, is key to finding water.

- **Understanding the Terrain:** Skip the guesswork and go for a systematic analysis of the terrain. Knowing how topography affects water movement gives you a much better chance of predicting where to find water.

- **Hydrological Know-How:** Get familiar with basic hydrology principles. That means understanding how surface water flows, how groundwater behaves, and how different soils handle water. This knowledge is your secret weapon.

Adapting to Different Environments

- **Tailor Your Approach:** Not all terrains are identical, so your approach shouldn't be either. Customize your methods based on where you are. What works in one place may not work in another.

- **Consider the Climate:** Think about how climate affects water availability. In dry areas, certain plants might be reliable indicators, while in wetter places, you may need to understand how water moves beneath the surface.

In a nutshell, let go of the myths and focus on solid, evidence-based methods. By understanding the unique challenges of different terrains and relying on science, you'll be much more successful in finding water where you need it. It's not just about survival—it's about being smart and informed in the great outdoors.

Chapter 3: Rainwater and Dew Collection

One of the first considerations you must make when pursuing a life off-grid is how you will get water. There isn't always the luxury of having a big body of water like a lake or river nearby, so rain and dew collection may be the only options for sustainable water collection. Any survivalist must have a constant supply of water, so collecting rain may arguably be the easiest way to achieve self-sufficiency.

When you open your faucet and water flows out, it has already gone through multiple processing steps to be safe for human use. If you collect your water, the responsibility for making sure your water is usable now falls fully onto your shoulders. Therefore, it is essential to learn the best methods of collecting water and how to process the life-sustaining liquid afterward.

Rainwater and dew.
https://pixabay.com/photos/raindrop-drops-car-roof-heaven-4929166/

By understanding the natural processes of nature and how you can use them to your advantage, you will be equipped with the knowledge to set up workable survival systems. Furthermore, you will be informed about the essential safety protocols needed to sustain your health and use water responsibly. From weather patterns and ecosystems to microbes and filtration, the world of rainwater and dew collecting is demystified with practical tips and credible information.

The Process of Rain and Dew Formation

To harvest dew or rain, you must be in the right environment and understand the functioning of natural processes. If you are in a drier area, you may need more storage, whereas if you are in a tropical climate where rain is abundant, the potential of flooding is going to be an issue to consider when collecting rainwater. If you are not near a natural water source, or the water body you have access to is contaminated, the best option you have is collecting rain and dew.

Rainfall is part of the water cycle, which has four main parts. The first section of the water cycle is evaporation. The sun heats surface water, speeding up the molecules so much that the liquid water turns into water vapor. The second part of the water cycle is when the vapor reaches the sky. The atmosphere cools the vapor, condensing it into clouds. When the clouds are heavy and saturated, the third step of the water cycle occurs, which is precipitation. Water vapor condenses back into liquid, causing it to fall back down to the earth. The last part of the cycle is the formation of groundwater, which is the best time to collect it. The cycle constantly repeats. You can collect rainwater at any stage in the cycle. A mesh net can be used to catch evaporating water. Fog catches can be used to gather low clouds in mountainous regions. The most apparent and efficient way to collect rainwater is to catch surface runoff.

Dew forms in a slightly different way than rain, but many of the same processes are still involved. Dew is a result of condensation, which is the process where gas transforms into liquid. At night or in the early morning, the surface of objects cools down because the sun is no longer hitting them directly. When a surface or object gets cold enough, the air around it will also cool down. Warm air holds water vapor better than cold air. The cold air condenses the water vapor, which causes small droplets to form. The dew point is the sweet spot where dew can develop. The formation of dew depends on numerous environmental factors like your geographical location, as well as temperature and wind. Dew forms better in humid regions like tropical rainforests. In deserts, the conditions are not suitable for a lot of dew to form because the air is too dry. Therefore, all you need to collect dew is the right surface and an appropriate region that has all the needed environmental conditions.

Rain and Dew Collection Methods

Since ancient times, humans have settled near water sources. If you study any civilization, you will find that they sprung forth from one centralized place that either had a river, spring, or lake. The Ancient Egyptian civilization was birthed from the rich agricultural land near the Nile, and Rome rose from a marshy swampland in Italy. When setting up an off-the-grid homestead, or if you are in a survival situation, sometimes the only option you have is to collect rainwater.

Rain and dew collection are cheap and convenient options for obtaining water, which can be set up quickly. There are some downsides of rain collecting, like seasonal rainfalls and droughts. However, with some luck and the proper setup, you could have a rain and dew water system that runs throughout the year. The easiest setup for collecting water is using tanks to collect surface runoff. Plastic water tanks created for this use are available on the market. The water tank will be set up near a roof or somewhere rainwater can flow off. Gutters are connected to the tank to collect the water. It's then stored in the tanks, which can be accessed when needed. The beauty of this method is that you can use the same principles to create cheaper options by using recycled materials. However, you should always ensure that the container you store your water in does not house harmful contaminants.

Smaller barrels can also be used as a cheaper option. However, the amount of water you can collect will drastically decrease compared to the much bigger water tank. The upside about using barrels is that they will be easier for you to move. Full water tanks need machinery to be relocated and will take far more effort. Setting up a few barrels around your home near the gutters can help you meet some of your basic water needs, like washing or watering the garden, but it is unlikely you can go fully off the grid by relying solely on this method.

A gabion is a more permanent way to collect rainwater.

A more permanent and labor-intensive way to collect rainwater is a gabion. This basic dam is made up of rocks placed within a wired boundary wall. A concrete mixture is then placed over the rocks to seal the gaps. You can collect a lot of rainwater depending on how big your gabion is. Since the water will be still, you must be precise with your disinfection and cleaning processes. The materials that are used are relatively low-cost. The wire mesh for the walls can be purchased at any local hardware store, and the rocks can also be bought or even collected depending on the land you have access to.

Swales and canals will be very useful if your rainwater is to be gathered for farming purposes. Swales are shallow ditches strategically dug to channel water from large surfaces downhill toward crops. If done correctly, this simple irrigation system can save you thousands of dollars per year in farming costs for any subsistence off-grid community or homestead. Many rural communities in Africa and Asia use swales because their livelihoods are directly linked to their success with the land.

Dew collection works best in areas with high humidity. Harvesting dew is an often-overlooked way to get water for self-sufficient living. Methods of dew collection can sometimes seem crude, but they are effective. The first simple way to harvest water from dew is to collect it from the grass. First, you will need a storage container. Then, you need an absorbent cloth made from wool, cotton, or a similar compound of synthetic materials. When the dew has settled in the early morning, lay your cloth on the grass to absorb the water. Once the cloth is heavy with moisture, wring it out into your water-collecting container. Repeat this process until you have collected all the water that you need. This method can also be used for other plants and leaves, but ensure your flora is non-toxic before applying this technique.

A waterproof tarp can be used for dew collection.
ModalPeak, CC BY-SA 4.0 <https://creativecommons.org/licenses/by-sa/4.0>, via Wikimedia Commons:
https://commons.wikimedia.org/wiki/File:Forester_Tent_Tarp_and_Poles_(6262CE06).jpg

A waterproof tarp can also be used for dew collection. When deciding upon any survival prepping or off-the-grid lifestyle, a central principle is to use what you have. Many people have tarps lying in their garages or basements. For this water-harvesting method, you could also use raincoats or an old tent as a makeshift tarp. Find a slope or construct one wherever you have some space out in nature or on your land. Use bricks and soil to pin the tarp down. Lay the tarp in a triangle shape on the ground, folding the sides over your bricks and soil and securing the material to the ground. Create a funnel shape at the end of the tarp.

Make sure that soil is packed underneath your tarp to cool the material at night. Place a container at the end of the funnel you have created and allow the dew droplets to flow slowly into it. Gravity helps you with this method. That's why you need to construct the contraption on a slope.

Traditional Approaches and Modern Innovations

Ancient cultures understood the value of rainwater collection. Thus, various systems were developed around the world to harvest rain. Today, rainwater is still collected in big cities through a complex network of gutters and greywater constructions that channel water into dams and through processing plants. Traditional rainwater harvesting methods are cruder than the marvels of the modern world, yet they paved the way for these new systems.

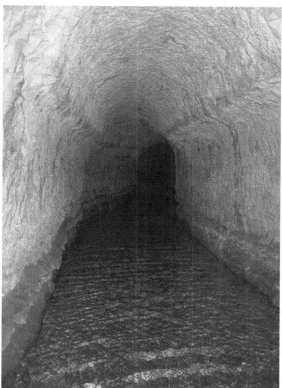

A qanat is a traditional rainwater harvesting technique used in the Middle East.
Pafnutius, CC BY-SA 3.0 <https://creativecommons.org/licenses/by-sa/3.0>, via Wikimedia Commons: https://commons.wikimedia.org/wiki/File:QanatFiraun.JPG

A traditional rainwater harvesting technique used in the Middle East is qanats. These engineering masterpieces channel water from slopes underground into an agricultural area or to be used for drinking water. The same method has been applied in various North African and Asian regions. In Oman, the construction is called a falaj, and in North African countries that used the technology, it was referred to either as a *khettara* or a *foggara*. In Central Asia and Pakistan, the term *karez* is used. All these names described the same basic structure. Vertical holes are dug into a slope, which all feed into an underground horizontal channel. The channel leads to the bottom of the slope, where the water is collected in a storage tank or a dam or dispersed into irrigation channels. Modern water collection techniques have replaced the need for widespread use of qanats. However, new interest in the technology is peaking as a potential method for sustainable groundwater collection.

Stepwells are a popular way of collecting rainwater in India.

In India and other areas in the subcontinent, the use of step-wells to collect rainwater was popular. Constructing a step-well is a commitment that requires many hands on deck. The structure takes up a large surface area to collect as much rainfall as possible. The spaces where step-wells were located were often communal, but recently, as technology and urbanization took place, they fell out of favor. A step-well is built in a square, with steps funneling the surface water to the narrow well in the center of the construction. Some of these old wells are elaborately decorated, showing how important they were to daily life for generations past.

Principles from these older structures can be applied to your personal water collection system, like using a large surface area to harvest water, manipulating gravity to your advantage, and constructing channels to transport your water underground on slopes. Some of these old principles have stood the test of time, but they can be combined with some modern innovations. The ancients were not only masters at rainwater collection, but they were also knowledgeable on multiple techniques of dew harvesting. In Israel, archeologists have discovered low circular walls built around vegetation to trap condensing water. South Americans and Egyptians used a similar method: pile up stones so that water vapor could cool in between the gaps under which a container was placed to catch the falling droplets.

It may be time to renew and improve some of these dew and rain harvesting techniques as multiple factors converge, making it a necessity in the future. With pollution and climate change, water will become scarcer, so individual families may need to find ways to collect rain. Furthermore, rainwater collection may create a more sustainable future as stormwater runoff is collected instead of letting the water flow, carrying pollution and contaminants. Moreover, surface runoff from water can cause soil erosion, which impacts food security. As financial instability increases all over the world, collecting rain and dew may also be a way to save money.

Underground storage tanks for water have proved to be a welcomed innovation due to the space that they save. In urban environments, space is a commodity that not many have access to. This problem has been solved by moving water storage underground so that more surface area can be available in cramped cities. Vertical rain gardens and green roofs have also gained popularity. Green roofs, which are also called living roofs, are gardens that are specially constructed on the top of a roof to grow flowers and produce. The soil acts as a sponge from which water can be collected. Moreover, green roofs can be energy-saving because they provide additional insulation in the winter. Vertical rain gardens are stacked plant setups placed near gutters to channel rainwater to produce food in limited spaces.

When it comes to dew harvesting, one of the central innovations in this field is the use of grooved surfaces. When grooved surfaces are used on a slope, the depth of the groove causes more condensation to occur, allowing for the collection of bigger droplets. This makes dew collection far more efficient and less labor-intensive. Fog collectors have also been introduced into some mountainous areas to collect low-lying clouds. The fog collector uses a mesh material made from synthetic fibers like nylon to catch droplets out of the atmosphere. These droplets are then channeled into plastic storage containers. Fog catchers only work in specific environments where it's moist and fog is common, but it is an easy and efficient way to collect moisture straight from the air.

Safety Considerations

Water collected from rain and dew is not immediately drinkable. Microbes and contaminants in the water could cause illness for humans and animals. You must be mindful of what you are going to use your rainwater for, like flushing toilets, washing, irrigation, or consumption. If you are consuming the water you collect, you have to be extra careful to make sure that it is drinkable.

Cleaning your water takes two steps, namely, filtration and disinfection. Filtering your water takes out all the bigger particles that may be harmful, like metals and dirt, while disinfection kills all the harmful microorganisms. There are different filtration methods that you can use. The filtration technique you use will determine how fine the particles that can go through your filter are. Cloth and sand can be used for filtration. Beach or river sand can be placed in a tube through which you allow your water to flow, or you can pour your water through some fabric. These filters are crude and should not be used for drinking or cooking water. You can purchase some great filters that attach to faucets and directly filter the water that flows from them.

To disinfect your water, you can either boil it or use chemicals. Boiling is an ancient, tried, and proven method that kills many microbes because they cannot survive in the harsh environment of those extreme temperatures. You should boil your water long enough to ensure that any harmful organisms are thoroughly taken care of. Using chemicals to disinfect your water is a little more complex because you must get the dosage right. Bleach and chlorine are two commonly used cleaning chemicals. However, they can be harmful for consumption if too much is used, so you must make sure you know the exact ratios you should mix. The taste and smell of chlorine can be removed from water by using specialized charcoal or carbon filters.

UV light is another way to disinfect your water. This method can be expensive and requires you to know how long your water must be exposed to the light and how much water one bulb can disinfect. This method is gaining popularity as UV disinfecting becomes more embraced socially. Distillation is another effective way to clean your water. Distillation is the process of boiling liquid until it evaporates, then recooling it to become liquid again. Distillation is expensive because you need a lot of energy to boil water to the point of evaporating it. It can also be used as a way to separate salt from water.

UV light can be used to disinfect your water.
Coleopter, CC BY-SA 4.0 <https://creativecommons.org/licenses/by-sa/4.0>, via Wikimedia Commons: https://commons.wikimedia.org/wiki/File:Uv_lamp.jpg

Before you go through the filtration and disinfection process, you must consider the equipment you use to collect rain and dew. Make sure that you keep your catchment areas clean and that you replace your filters regularly according to the manufacturers' guidelines. If you are using roofs as water catchment spaces, make sure that you clean them properly before rainfall or big storms because a lot of dirt and germs can gather on them when exposed to the elements. Lastly, remember to keep your clean water and the liquid you have just collected separately.

Your tanks and storage barrels should also be made from materials that do not contaminate the water and should be cleaned regularly. If your tank is empty, run the first water collected in it out because it is likely to contain dirt that has piled up at the bottom.

Chapter 4: Rainwater and Dew Storage

Since collecting rainwater and dew are easy and cheap options for obtaining water, many people gravitate toward these methods. However, gathering the liquid is just the first part. Safely storing water and ensuring that it is potable could mean the difference between life and death. There are hundreds of waterborne diseases with excruciating symptoms that stem from consuming untreated, stagnant water or as a result of weak safety protocols. Your well-being is in your hands because you are solely responsible for finding out how to store water appropriately. With a basic understanding of the considerations you need to make when storing water, you can take the necessary steps to keep clean water over long periods.

Safely storing water is paramount to the success of your endeavor to survive independently.
https://pixabay.com/photos/metal-container-technology-8332370/

The containers you use, as well as your disinfecting and filtration techniques, play a role in storing a safe, drinkable supply of water. Your maintenance routine and the care you take in keeping your water contaminant-free will hugely impact your health. In an outdoor survival situation or on an off-grid homestead, safely storing water is paramount to the success of your endeavor to survive independently. By learning the details of what makes water safe to drink and what steps you can take to maintain your storage setup, you can thrive with an abundance of potable water.

Long Term Storage Containers

Unless you are living in a tropical region where it constantly rains, the wet season will pass by quickly. This means there is a limited window for you to gather as much water as possible. The way you store your water will impact how usable it is in the dry seasons. There is no expiration date on water, so you can store it as long as you need to. However, water is a carrier for all kinds of microbes, parasites, and contaminants. The containers you use to store your water are the first line of defense against disease-spreading microorganisms, so you'll need to put a lot of thought and research into the containers you use and not store them recklessly by pouring water into any plastic bottle you find lying around.

When you think about how rainwater is collected, it is easy to see how it can become dangerous for humans to consume. For example, if your tank is connected to gutters from your roof, bird excrement can be washed into your water supply, and no one would drink a glass of water with fecal matter floating around in it. You must think about the array of foreign particles that could wash into your water supply. Although collecting rainwater is convenient, it is not pure by any means. Rain can gather all kinds of dangerous pathogens from the surfaces that it touches.

The first stop when considering long-term water storage is your containers. The tanks and barrels that you keep water in must be food-grade. All plastics are not created equal. There are many different ways to manufacture plastics that contribute to the quality of the material. Food-grade plastic is of higher quality than containers not intended to hold water. If you look at the plastic containers that you use, there should be a number in a triangle. This number is called the resin identification code, or recycling number, and it tells you what kind of plastic your container is made from. Many food-grade plastics will have the code HDPE or LLDPE, but this way is not the most reliable when determining if containers are safe for storage. The same kind of plastic that is used for water can also be applied in other instances, like for fuel, so be careful because it is doubtful that you want gas-flavored water. Safe RIC codes that you can store water in also include PETE and PET.

The FDA determines which plastics are safe for water and food to be stored in. Plastic does tend to leak into your water, so it is not 100 percent safe. However, some plastics are safer than others. Glass is probably the safest material to store water in. Still, it is impractical because there is no glass container big enough to carry huge supplies of water, and the containers are easy to break, so an extra level of care must be taken. However, a small glass container may be beneficial if you are outdoors in a wilderness survival situation, but there is still the risk of breaking the fragile material.

How well your containers are sealed also affects the contaminants that can enter your storage vessels. Your containers should be filled to the top so that there is no air in them. Moreover, your containers should be clearly labeled with what the water can be used for and the date it was stored. You could use some water for plants and flushing toilets, but it may not be appropriate for consumption. To avoid mixing up the uses for the different water you have stored, keeping them separated and clearly marking them is essential.

The water you store should be shaken or agitated at least every six months if not used. This may be enough time for the new rainy season to come back, but it isn't advisable to drink that water anymore if it is not. Drinking the water you have been using without incident throughout the year can be tempting, but all it takes is one slip-up for disaster to strike. You should be as careful with your water as you would be with a fragile newborn baby.

Some people have used canning jars for drinking and cooking water, which is also a brilliant solution for fighting microbes and reducing contaminants. These jars are easy to store and can be packed in boxes until needed. Since glass is the best material for storing water, tightly sealed jars are ideal. Furthermore, using canning jars also allows you to transport your water easily. Make sure that your jars are clean before you pour any water into them. Another benefit of using jars is that you can differentiate water meant for other uses and water meant for consumption because all your drinking water will be preserved in these sealed containers instead of in large plastic tanks.

Keeping Stored Water Free from External Contaminants

Getting the right containers is only one part of the battle against illnesses that come from water. Once your water is in your chosen containers, you still have to make sure that no external contaminants infiltrate your tanks, barrels, or jars. An air-tight seal and a container filled to the top are essential. Your water should also be kept in a cool, dry place that is between 50- and 70 degrees Fahrenheit.

Just like you must clean the surfaces from which your water is collected, you should also sterilize your containers. First, wash your vessels with warm water and soap. After you have rinsed all the soapy water out of the container, it is time for disinfection. Mix one teaspoon of unscented chlorine bleach to about four cups of water. Your bleach should range from 5 percent to 9 percent sodium hypochlorite. Shake the container well and let the mixture rest for about thirty seconds. After that, you should rinse the container thoroughly and let it air dry. This will eliminate any parasites or pathogenic bacteria that may have gathered inside the vessel.

If you store water for long-term use, it is likely that your water will be kept in bigger containers and then decanted into smaller ones. Any utensil you use to transfer your water from one vessel to the next should also be kept clean and sterilized. If you dip a dirty bucket into a water tank, the hundreds of liters in that tank are now also contaminated. Therefore, it is also advisable to never let your hands come into contact with your water supply because numerous pathogens are living on your skin. Humans can be a walking disease factory, so be mindful of your body when you interact with your water.

Your environmentalist leanings may prompt you to use recycled containers to store your water. This is fine as long as no toxins were previously stored in the vessel and the material is food-grade. However, you also need to ensure that there are no cracks or holes in your recycled containers. A damaged container can easily become a superhighway for disease. Inspect your containers before using them to ensure that they are in peak condition, especially if they are going to be used for long-term storage. This is also why you must seal your containers properly. Another helpful tip is to use some storage containers with narrow necks so you can pour the water out without touching any of the contents.

Some basic hygiene protocols are crucial when working with water, like washing your hands before handling any of your storage containers. Keep the environment where you store your water clean as well. It is better to elevate your storage to avoid getting contaminated with dirt or animal waste. Hygiene is key because you are fighting invisible enemies like parasites, protozoa, and pathogenic bacteria.

Some people use a pond to store their water.
https://unsplash.com/photos/a-pond-surrounded-by-trees-in-the-middle-of-a-forest-7CdNNvijw9E?utm_content=creditShareLink&utm_medium=referral&utm_source=unsplash

Some people may use a pond or a dam to store their water, but this method is more dangerous because your water is exposed to the elements. It is not advisable to drink water that has been stored openly in a dam. There may be other uses for this water, like watering crops or flushing toilets. If you have water that is being stored in the open like this, you must be extra careful to prevent it from mixing with your potable water. Water in dams tends to be cloudy, so it is the perfect breeding environment for microbes and living organisms like mosquitoes and mollusks, which also call stagnant ponds or dams home. Their waste could be dangerous for humans, not to mention that mosquitoes in some regions carry malaria.

Dangers of Water Stagnation and How to Prevent It

If you've ever watched a survival video or television program, you'll notice that when people are looking for water, they never drink from a stagnant source and opt to seek out flowing water like a river or a waterfall. Stagnant water is alive, but not in a good way. Insect larvae, parasites, and other microbes make their home in stagnant water. This is concerning because if your water is being stored in barrels, tanks, or other containers, there is a good chance that it is not flowing. Therefore, there are safety steps you must take to maintain your well-being when consuming the rainwater you have stored.

Stagnant water is dangerous for human consumption.
Bibiire1, CC BY-SA 4.0 <https://creativecommons.org/licenses/by-sa/4.0>, via Wikimedia Commons: https://commons.wikimedia.org/wiki/File:A_stagnant_water_at_Ilepo_Araromi.jpg

Stagnant water is categorized as the most dangerous class of water for human consumption. Water is classed into four categories, namely, clean water, grey water, slightly contaminated water, and black water. Stagnant water outside falls under the class of black water because the pathogens that it breeds are so dangerous to humans. Some of the deadliest diseases and parasites with horrific health costs, like salmonella and E. coli, are found in stagnant water. Viruses like hepatitis E and rotavirus also thrive in stagnant water. The mold that is produced in this black water can be dangerous. The combination of bacteria, viruses, parasites, and mold makes stagnant water something that you do not want anywhere near your property, especially close to your drinking water.

It is common for people to pump stagnant water into their plants, but experts recommend that it should be contained and expelled completely from a property. Mounted trucks can be called in to get rid of this water. Willingly introducing stagnant water onto your property is illogical. Store your water in sealed containers to prevent the laundry list of issues related to stagnant water.

Sealed bottles with narrow necks are your best option to prevent the contamination of water that isn't flowing. Even in this ideal situation, you must swap out your water every six months to a year. The tight cap will prevent unwanted visitors, and the vessels should be kept out of direct sunlight. Suppose your water is being stored outside in a dam you have constructed. In that case, you must take the necessary action to get your water flowing. You'll notice that algae starts to form if water is still for a long enough time. This is a bad sign that your water is not potable and can be dangerous if handled. There are portable pumps available that can be used to circulate the water in your storage dam. You will have to factor in the cost of electricity or fuel needed to power these pumps. The bigger your water reserves are, the more powerful the pump you need will be. Circulating the water supply for a few hours a day should be enough to prevent algae buildup, but it does not combat other microorganisms that are not as easy to see.

If you are not drinking water stored in an outside dam, you may be tempted to let the algae develop. However, there are many dangers associated with algal growth. It can clog up your pipes if you have irrigation systems and rust metal components on your homestead; it can make your animals sick and even kill some plants. Preventing algae from forming is a better option than looking for ways to treat it afterward because some of the algae eradication chemicals also contaminate your water.

Big storage tanks sometimes promote algal growth. Keep it out of the sun in a cool and dry place to stop it from growing in your tanks. You also need to flush your tanks, especially after the first rain, to get all the access debris that piles up in the bottom out. Algae will not grow in a properly sealed tank because you are depriving it of the air it needs to grow and spread. Shallow and still water is where algae thrive, so if you construct a dam, it takes extra effort and resources to dig it deep.

Maintenance Routine to Keep Water Potable

Like any machinery you run, your rainwater storage and distribution must be maintained. Everything may seem to be working well, but all it takes is one overlooked detail for your system to spiral out of control. The consequences of not keeping up with maintenance on water gathering and storage can be horrific. Illness and diseases can creep in if you don't keep on top of your repairs.

The first check that you should make is that your containers are not damaged and that they seal properly. Contaminants slip through the smallest cracks, so you must constantly check for leaks. Replace tanks or barrels leaking so your storage systems run at peak conditions. Some plastic containers have expiration dates on them, so if you are using water bottles for storage, you must check that your vessels are still up to date.

Cleanliness is the foundation of keeping water potable. Your runoff surfaces should be kept tidy, especially before it rains. Furthermore, big water tanks gather dirt and debris at the bottom, so you need to flush out these containers occasionally so they don't become attractive to pathogens. Any pipes, gutters, and fittings attached to your tank should also be spotless.

Even when you keep water in food-grade sealed bottles, plastic will leak into the water after a while, so rotating your water every six months is essential if you are keeping an emergency supply. Ensure your storage room is also disinfected with bleach, and your water bottles are kept off the floor. Filtration is a must, but your filters should also be changed regularly according to manufacturing guidelines. Filtering removes various parasites and debris from your water. Only drink clear water that has been filtered and disinfected so that you do not get any viral infections. Your containers must be kept clean, so remember to disinfect your vessels after each use.

Filtration and disinfection are aspects which must be taken seriously and which cannot be neglected. There are many disinfectants on the market specifically for drinking water. Chlorine and iodine tablets are available to clean your water, but remember that these methods should only be used on clear water and not murky water because some microbes can hide within dirt particles to avoid the disinfecting effects of these chemicals.

Keep all your clean water for drinking and cooking away from water meant for other uses. The water you use for your toilet may not be as clean as the water you consume. Labeling is an essential part of your maintenance routine, so you should replace any damaged labels and mark new containers you introduce into your water storage system. Transferring your water from one container to another regularly also helps prevent the water from tasting bad. Your hands and clothing should be clean when handling your drinking water, so institute cleaning protocols before you enter the area where you store your water.

Chapter 5: The Art of Water Purification

After collecting water, you may be tempted to drink or use it right away. However, rainwater is contaminated and contains germs and chemicals that can be harmful to you and your family. Purifying the water will eliminate all toxins and render the water safe for consumption. There are many purification methods that you can use; some are primitive, while others are more advanced.

This chapter covers the inherent risks of untreated water and different purification techniques with step-by-step instructions for each one.

Purifying the water will eliminate all toxins.
https://pixabay.com/photos/water-drops-falling-droplets-20044/

Risks of Untreated Water

Untreated water is water that hasn't been purified or filtered and contains sediments, chemicals, and microorganisms like parasites, viruses, and bacteria. Consuming this water can cause various health conditions and, in severe cases, death.

Health issues caused by untreated water:

- Headaches
- Fatigue
- Fever
- Polio
- Typhoid
- Hepatitis A
- Fluorosis
- Dysentery
- Diarrhea
- Kidney and liver damage
- Cholera
- Gastrointestinal problems
- Neurological problems
- Stomach aches
- Lead poisoning
- Dehydration
- Amoebiasis
- Arsenicosis
- Nausea
- Salmonella
- Trachoma (eye infection).
- Poliovirus
- Cryptosporidiosis
- Cancer

Drinking untreated water is particularly harmful for people with weak immune systems, the elderly, children, and infants. They are more likely to get sick and may even die from the chemicals and microorganisms.

Protect your family's health by purifying the water with any of the following techniques.

Boiling Water

Boiling is one of the oldest and simplest ways to purify water. Older generations have been relying on this method for centuries. Even with the invention of filters and mineral water, some still prefer to use it. You can use this technique anywhere since it is fast and doesn't require any equipment.

Boiling is one of the simplest ways to purify water.

Advantages of Boiling Water

- It is one of the quickest and most convenient methods. You can boil water in a kettle, microwave, stove, solar cooker, gas stove, gas grill, fire pit, wood stove, or survival stove.
- It is cost-effective.
- It improves the water's taste by removing undesirable odors and flavors.
- It kills viruses and bacteria and removes solids.

Disadvantages of Boiling Water
- In some cases, it changes the taste, giving it a strange flavor.
- It requires heat and energy, which can be inconvenient in regions where energy is expensive and heat is unavailable.
- It is time-consuming, especially if you are boiling large amounts of water.
- Boiling water doesn't eliminate many harmful substances like lead, chlorine, microorganisms, heavy metals, chemicals, hormones, fertilizers, pesticides, and microplastics.

Instructions:
1. Put the water in a big stainless steel pot.
2. Put it on any source of heat available.
3. Leave it for three minutes to boil.
4. Then let it cool down.
5. Store it in a clean and sanitized container and seal it tightly.

Clay Vessel Filtration

Clay vessel filtration is one of the world's oldest and most traditional water purification methods. Before using advanced methods like reverse osmosis and UV radiation, people treated their water in clay pots. Clay works by trapping mud, contaminants, and impurities, leaving the water safe. Although it is a primitive method, there are still people using it to this day.

Advantages of Clay Vessel Filtration
- Eliminates impurities, bacteria, and protozoa in water
- Improves the taste
- Enriches the water with healthy minerals like iron, magnesium, and calcium
- Environment-friendly
- Cost-effective
- Accessible and easy to use

Disadvantages of Clay Vessel Filtration
- Doesn't remove viruses from water
- The water is prone to recontamination

• Requires regular cleaning

Instructions:

1. Pour the contaminated water into a clay pot
2. Cover it with its plastic lid
3. Leave the water to filter
4. You can drink it from the clay container

Desalination

Desalination involves removing salts and minerals from seawater or any other type of saltwater. It makes water with high levels of saline safe for consumption. Although it is more common for commercial use, like on boats, in resorts, hotels, and residences, it can be suitable for individual use. This method works in areas that only have access to saltwater.

Desalination removes salt and minerals from saltwater.

Advantages of Desalination

• Provides clean and fresh water
• Removes salt from water
• Eliminates toxic minerals and compounds
• Separates microorganisms from the water

Disadvantages of Desalination
- Requires a lot of electricity (but not for individual use)
- High-cost

Instructions:
1. Get a large bowl for the salt water and a container for the condensed water. Make sure the bowl is much bigger than the container.
2. Fill the large bowl with salt water and put the small container in the bowl. It should be floating on it, and you must cover it with a paper bag.
3. Place them under direct sunlight, like outside or near a window.
4. Put a small rock on top of the paper bag to push the condensed and treated water into the cup.
5. Leave them for four hours.
6. You should find fresh water in the container. This is the treated water that is safe for use.

Distillation

Distillation is a water purification technique that involves the use of steam and condensation by boiling the water and converting the steam to liquid. The process eliminates any microbes in the water but doesn't remove metals, solids, and some pollutants. It is ideal for purifying large amounts of water but requires heat or solar energy.

Advantages of Distillation
- Perfect for people who live off-grid since it doesn't require the use of electricity
- Produces clean and high-quality water
- Removes salt from the water, making it safe for consumption
- One of the safest water purification methods
- It doesn't use filters or chemicals
- Removes bacteria, fluoride, and heavy metals

Disadvantages of Distillation
- Takes longer than other methods
- It isn't cost-effective

• Quality may vary depending on the equipment you use

Instructions:

1. Put the water in a large pot and put an empty small pot inside it.
2. Leave the water to boil on a stove or under the sun.
3. Leave it on medium heat.
4. Place the lid upside down over the large pot to allow the condensed water to trickle down into the small pot.
5. Then, place ice on the flipped lid. The difference in temperature on both sides of the lids will increase the speed of the condensation process.
6. When the ice melts, add more.
7. This will take some time, depending on the amount of water you use.
8. You will know it's done when the large pot is empty and the small one is filled with the treated water.

Iodine

This method is a little controversial since you are trying to remove chemicals from the water and not add more. However, life off-grid and emergency situations won't always give you a lot of options. So, only use this method when you can't use any of the other techniques. The best chemicals for treating water are iodine and chlorine.

Iodine tablets can eliminate viruses and bacteria from water.
Mx. Granger, CC0, via Wikimedia Commons:
https://commons.wikimedia.org/wiki/File:Iodine_pills.jpg

Iodine is a red chemical that is sold as a liquid or in tablet form. It eliminates viruses and bacteria from water but leaves an unpleasant taste. If used in high dosage, iodine can be fatal. So be extremely careful when using it, and it should only be a last resort.

Advantages of Iodine

- It is convenient and cost-effective
- It kills protozoa, viruses, and bacteria
- Lightweight
- Easy to use

Disadvantages of Iodine

- Strong aftertaste
- Not safe for pregnant women
- Fatal in high dosages

Instructions:

1. Add five drops of 2% tincture of iodine to one liter of water and 10 drops if the water is colored or cloudy.
2. Stir the water, then leave it for 30 minutes to an hour before using it.

Ion Exchange

This technique works on removing arsenic, nitrates, radium, and barium from the water through ion exchange. The great thing about it is that it treats water without affecting its taste. This technique is very powerful since it removes heavy metals that many other methods fail to eliminate.

Advantages of Ion Exchange

- You can use it with other techniques
- It can be used on a large scale
- Easy to use
- It is safer than many other water treatment techniques
- Removes bad odors and tastes from the water
- Easy to install
- Easy to maintain

Disadvantages of Ion Exchange

- The stored water has a short shelf life
- Expensive
- Uses chemicals like salt
- Can't eliminate bacteria, viruses, pyrogens, or particles

Types of Ion Exchange

- **Water Softening:** It removes minerals that harden the water, like magnesium and calcium, and replaces them with sodium ions to soften the water.
- **Deionization:** It removes anions, cations, and all ions, resulting in purified and uncontaminated water.

Instructions:

1. Install an ion exchange water filtration system.
2. Once installed, the water will be safe for use, and you can consume it right away.

Reverse Osmosis

Reverse osmosis (RO) is a water purification technique that separates dissolved solutes from water. It eliminates heavy metals, microplastics, PFAS, VOCs, arsenic, chlorine, sediment, salt, fluoride, herbicides and pesticides, unwanted molecules, ions, and other contaminants. It is one of the most popular methods of purifying drinking water. People have been using this method for decades, and it is considered one of the most significant inventions in modern history.

Advantages of Reverse Osmosis

- Removes chemicals, germs, bacteria, viruses, and other biological contaminants
- Small and easy to use
- Doesn't require electricity or energy usage
- Improves water waste
- Fully automotive
- Removes impurities like fluoride, chlorine, and lead
- It does not use any chemicals to purify the water
- It is extremely popular in commercial use because it is adaptable and can cater to several needs

- Suitable for private and public use
- Produces high-quality water

Disadvantages of Reverse Osmosis
- Produces a lot of waste
- It is costly and requires regular maintenance since it can be clogged
- Eliminates all types of minerals, including the good ones your body needs
- Using it regularly can sometimes lead to cardiovascular issues, weakness, muscle tension, and fatigue
- Doesn't remove microorganisms like bacteria and viruses
- Doesn't disinfect the water
- Requires more energy than other methods
- Produces waste

Instructions:
1. Install the reverse osmosis system in your home or RV under the sink or outside. Or you can hire someone to do it for you.
2. This is an automatic method that will purify the water by itself.

Shungite for Water Purification

Shungite water purification isn't as popular or commonly used as the other methods in this chapter, but it is a very interesting one. Shungite is a rare mineral that you can only find in Russia. Some scientists believe it came from a meteor that crashed into Earth centuries ago. It contains high levels of fullerene carbon, a rare carbon structure that can absorb contaminants from the water.

Advantages of Shungite Water Purification
- It has antiviral and antibacterial properties
- Destroys viruses and bacteria from water
- Kills microbes in water
- Removes dangerous pathogens that are considered health hazards
- Purifies water
- Protects against EMF exposure

Disadvantages of Shungite Water Purification

- It can release chemicals and heavy metals into the water
- This method is a little risky, so caution is advised

Instructions:

1. Get one or more Shungite stones.
2. Wash them thoroughly for two minutes under water.
3. Then, wash them again five times.
4. Soak the stones in the water and leave them for five days.
5. Change the water every day.
6. Remove the Shungite from the water, put it in a container, and then cover it with the water you want to purify.
7. Leave it for 24 hours to seep.
8. Then remove the stones and drink the water.

Solar Purification

This technique treats water using UV radiation. Solar energy removes contaminants from water, reduces microorganisms, and prevents them from reproducing, making the water safe for consumption. It is one of the simplest methods you can use. However, it won't work in cold regions or places without direct sunlight access.

Solar purification isn't as powerful as other techniques, so use it only when you don't have other options.

Advantages of Solar Purification

- Removes microbes from water
- Improves water quality
- Environmentally friendly
- Cost-effective

Disadvantages of Solar Purification

- Doesn't work in cool regions or cloudy weather
- Has no impact on extremely contaminated water
- Doesn't remove chemicals from water
- Only works with small amounts of water

Instructions:

1. Fill a plastic bottle with water.
2. Shake it to activate the oxygen.
3. Place it horizontally under direct sunlight.
4. Leave it for an hour.

UV Radiation

In this method, you use UV bulbs or lamps that emit UV light that can eliminate some microorganisms. However, this method isn't strong enough and can't remove heavy metals or impurities.

Advantages of UV Radiation

- Kills protozoa, bacteria, viruses, and microbes
- Environment-friendly
- Easy to install and use
- Requires little maintenance
- Doesn't use much electricity
- Doesn't waste water
- Water is instantly purified, so you can consume it right away
- Cost-effective

Disadvantages of UV Radiation

- It isn't effective against all microorganisms
- You will need to filter the water before using this method
- Requires electricity, which isn't always available for people living off-grid
- Doesn't enhance water's taste or odor
- Doesn't remove metals or solids

Instructions:

1. Put the water in a UV water purification machine and let it do its job.
2. The machine exposes the water to UV light, which kills microorganisms' DNA and prevents them from reproducing, making the water safe for use.

Water Chlorination

Chlorine is the second chemical you can use in treating water. Water chlorination is one of the oldest purification techniques. It is ideal to use during emergencies since it is simple, fast, and effective. It involves adding a mild bleach with 5% chlorine to the water to remove toxins and microorganisms.

Advantages of Water Chlorination

- Deactivates microbes and eliminates harmful microorganisms
- Disinfects the water, making it safe for consumption
- It is easy to use and brings fast results
- It is cost-effective and doesn't require any equipment

Disadvantages of Water Chlorination

- Leaves a strange odor in the water
- Constant exposure to chlorine-treated water dries and weakens the hair
- Causes skin irritation
- Chlorine changes the water's taste

Instructions:

1. You will need chlorine bleach to disinfect the water. Choose plain liquid laundry bleach like Purex or Clorox.
2. Check the ingredients and avoid using bleach with scents and additives.
3. Determine the amount you will need in advance. Use it moderately to avoid any issues.
4. Use four drops of chlorine for every 10 mils of water, or 8 drops if the water is very contaminated or cloudy.
5. Add the chlorine to the water and mix them well together.
6. Leave the chlorine water for 6 to 12 hours. Don't use the water before this time.
7. Test the water for chlorine using a digital meter.
8. Make sure the water is clean and chlorine-free before using.

Water Purifier

Water purifiers eliminate water impurities, minerals, and biological contaminants. Although they require electricity, there are many options on the market that you can use off-grid, like a pump water purifier. This is one of the most effective methods since it removes more contaminants than other techniques. However, it still doesn't remove chlorine and lead.

Advantages of Water Purifiers

- Environment-friendly
- Cost-effective
- Improves odor and taste
- Improves health
- Provides safe and clean water
- Kills microorganisms
- Easy to use

Disadvantages of Water Purifiers

- Water purifiers require maintenance
- They don't remove pesticides

Instructions:

1. Install the pump water purifier using the instructions on the package.
2. Fill it with water.
3. Pump the handle up and down for a couple of minutes.

Traditional vs. Advanced Water Purification Methods

Now that you have familiarized yourself with different purification methods, you are probably wondering which one is right for you. The answer to this question depends on many factors, such as the area you live in and if you live in a cabin, house, or RV. So, consider all these factors before making a choice.

However, the advanced ones take the lead when comparing traditional to advanced methods. Thanks to technology, there are newer, safer, and better methods to purify water. For instance, a traditional method like boiling water isn't always safe. Although your grandmother would argue that her family has used it for hundreds of years, recent discoveries have shown its limitations.

Things have changed in the last few years, and there are new types of viruses, bacteria, and heavy metals that traditional methods struggle against.

Advanced methods are better in every way than their traditional counterparts; although they may be costly and some can be less convenient, they are the safer option. You use water in every part of your life, so you shouldn't take any risks.

There is no denying that living off-grid has its challenges, and during an emergency, safety can be a luxury you can't afford. In this situation, you will use what is available to you rather than what is safer. However, if you can be prepared in advance, always choose safety.

Water purification isn't optional. It is a necessary process to remove chemicals and microorganisms from your water. Make sure to treat the water right after you collect it using the appropriate method. Stick to the instructions and don't change or ignore any of the steps, especially with chemical doses.

What would you do if you lived in a cold environment where all the water sources were frozen? The next chapter provides tips and skills to turn snow and ice into water without risking your health.

Chapter 6: Snow and Ice: Melting Tips, Myths and Misconceptions

In this chapter, you will learn all about carefully sourcing water from ice and snow, as it is often the only lifeline in colder areas. This chapter will equip you with the knowledge and skills to make informed decisions when sourcing water from ice. You will learn all about the energy conservation techniques for melting snow and about the debunking of related myths and misconceptions. You will also be learning about the proper way of treating ice while avoiding the risk of hypothermia. Let's dive right into it:

Turning snow into drinking water to keep yourself hydrated is an essential skill to master if you live in cold environments.

https://pixabay.com/photos/ice-melt-frost-melting-frozen-570500/

Efficient and Energy-Conserving Techniques for Melting Ice and Snow

Turning snow into drinking water to keep yourself hydrated is an essential skill to master in cold environments. It is also an essential skill that hikers and mountaineers must also learn. You can eat or drink snow by filtering it properly. You can consume a small amount of ice or snow for survival. However, ingesting large quantities can be extremely dangerous.

Consuming snow can lower your body temperature as your body requires a significant amount of energy to convert ice and snow into water, and the energy used causes your body temperature to change. You must keep your body temperature high, especially if you're hiking long distances. This is why it is critical to use efficient ways to convert ice to water. Let's take a look at some of them below:

Use a Stove

Many outside venturers travel in ice-cold environments with their modern stoves that require liquid fuels such as Coleman fuel, propane gas, or unleaded petrol to melt ice and snow. Don't fill your pot to the top with snow. It's a great insulator and may cause your pan to burn rather than melt the ice. The most efficient way to do it is by collecting the snow from outside your campsite and then melting a small amount of it in your pan on a gentle flame (Ensure that the snow you're collecting is not yellow). Add small amounts of snow to the pot until you have enough water. The snow mainly comprises air, so it takes a significant amount to produce a liter of water. To save fuel, you can also cover the pot or pan with a lid. You can also heat it over a fire if you have trees in your surroundings. The water from snow is not safe just because it is frozen. Bacteria and pathogens may become dormant when frozen but will be reactivated when the ice melts. Hence, it is preferred that you boil your water before consumption. Practice the same discretion with standing water.

Heat It Over a Fire

You can also melt your snow over a fire if you do not have a stove or a pan. You may hang the ice and snow over the fire and let the melted water drip into a container. You may wrap the ice in a cloth, a shirt, or any other porous item that can be hung with the support of sticks or even ski poles. You can also heat a hard sheet of ice or crusty snow over a fire without using any vessel. You may cut the sheet of ice into smaller pieces and put it over the stick to melt it.

Use the Sunlight to Your Advantage

It is alright if you do not have access to fuel or fire; it's still possible to melt ice without fire. You can also use the sun to melt your ice. When the dark rocks come in contact with sunlight, it becomes easier to melt ice or snow that comes in contact with them. You can find water even in temperatures as low as -10 degrees Celsius. Listen out for flowing water on cliff faces or south-facing slopes. Once you find your water, you can collect it in a container. It is important to find it and collect it before dark, or it may freeze again once the sun goes down. You can keep your water bottles in your sleeping bag to prevent them from freezing again.

Use Body Heat

If you carry a water bottle, keep it in contact with your body or inside your coat to keep the water from freezing. Take some snow and put it in your bottle whenever you sip your water. Your body heat (and the water in your bottle) will melt it quickly and replenish your water supply. Ensure you keep a wide-mouthed bottle so it is easier to put snow in it. This is one of the easiest methods to get water and does not require you to make a fire. It also works in emergencies or bad weather conditions.

Use a Zip-Loc

Another way to melt the snow under the sun is by putting it over a garbage bag called a drum liner. You place your snow in a Ziploc and then place it over the garbage bag. The bright heat from the sun will melt the snow inside the Ziploc and give you clean drinking water. This method works best if you are staying in areas with a bright sun. However, if you are not carrying Ziplocs, you can just place the snow on the drum liner and wait for it to melt. You can then put the water in a container. This process is difficult, and collecting the melted water this way may become a hassle.

Melt Ice on a Tilted Rock

This is another easy way to melt ice. You can melt snow with some rocks and a fire. Create a tilted table-like surface to hold your ice so that you can heat from below the rocks with fire. Then, channel that melted snow into a container below the rocks. You must use clean snow so you don't need to filter the water. It is important to use rocks that are not porous. Otherwise, the water will soak into the rocks, and you will lose it. You should also avoid taking the rocks from near the river, as they are more likely to become highly saturated with water. Suppose you are in areas like Coastal Alaska or British Columbia. In that case, you will be better off melting the sea ice that is found at a distance from the coast rather than the recent ice.

Eating Snow Directly

This should be the last resort, as eating a lot of snow can dangerously lower your body temperature, which increases the risk of hypothermia. However, very cold ice can burn your mouth and lips.

Melting Tips

It requires a lot of energy (i.e., fuel) to melt snow, as snow melts slower than ice, and fuel will probably be scarce. This is why poor melting methods can lead to excessive use of fuel. Here are a few melting tips that can help you conserve energy:

Protect Your Stove from the Wind

You must find a sheltered place for your stove as the wind can increase the amount of fuel that it takes to boil a liter of water. If you have a lot of snow around you, then try cutting snow blocks to create a windbreak for your stove. Although it isn't perfect, it will still protect your stove from wind. You may also use bushes or piled rocks to protect your stove from the wind. You can then use your own body for extra protection, too.

Heat with Water in Your Pot

Melting snow in a dry pot can be a slow process. You should always try to keep a little bit of water in your pot for efficient energy transfer to the snow. First, add an inch of water to the pot and then put in the snow. The snow will melt, and the water level will rise. Then, pour out the water from the pot while keeping the same amount, one inch of water at the bottom of the container.

Do Not Heat the Water

If you aim to just melt snow and not heat water, you must not waste energy heating water. Keep removing the liquid water from the pot as it melts while adding snow to the pot. You can add snow and stir the pot before you take out the water, as it helps recover the energy lost while heating water.

Use a Lid

Putting a lid over your pan or patting while melting the snow is a great way to conserve energy. It is quicker as well compared to melting the snow without a lid. If you put a lid on your pot, you trap the hot air and steam inside the pot, which fastens the melting process and significantly enhances the melting efficiency. It also reduces the risk of spillage while stirring the spot.

Establish Redundancies

If you are going on an outdoor adventure, ensure you have a working stove, fuel, and a pan, as these will make it easier for you to source water. If you do not have this equipment, there are other passive ways to melt snow, as discussed above.

Debunking Myths and Misconceptions

Although snow can be a wonderfully convenient water source, it is important to debunk some myths and misconceptions related to it. It is important to clarify these misconceptions for your and your loved ones' safety. Being aware of these issues helps you make informed decisions. Let's jump right into it:

Myth #1: Eating Snow Can Hydrate You

Reality: Eating snow can be great for extreme survival conditions, but you should not rely on it fully. People tend to misunderstand that snow is made just out of water, so it can be a great source of hydration. However, it is not as simple as it seems. Although snow is made of water, many good properties are lost in the water-freezing process, which is why eating snow can still result in dehydration instead of preventing it, as the body uses more heat and energy to melt the cold snow. This can lead to a reduction in your body's fluid level while making you more dehydrated than before. It also does not contain any essential nutrients and minerals you will need for proper hydration. However, melted snow is a great water source in emergencies and if you're outdoors with no access to streams.

Myth #2: Snow Is Purified Upon Melting

Reality: Snow does not automatically purify upon melting. Although it does remove some impurities, it does not eliminate all contaminants. Boiling the melted snow or using other water purification methods is the best way to remove harmful chemicals, pollutants, and microorganisms.

Myth #3: Snow and Ice Are Not Equivalent

Reality: While it is true that snow and ice can be melted to yield water, their processes may be different. Snow is composed of air, so it takes more time for the snow to melt than dense ice. This is why you should be patient and take care to melt ice and snow in the correct way.

Myth #4: All Methods of Melting Snow Are Equally Safe

Reality: All methods of melting snow may seem equally safe, but that is not true. If you try to melt snow in a plastic container or a contaminated container over an open flame, harmful pathogens and contaminants may be introduced during heating. Using a camping stove and a clean container is highly recommended.

Myth #4: Yellow Snow Is Safe to be Melted and Consumed

Reality: Yellow snow can be quite dangerous to consume as it may contain impurities and contaminants such as urine. It would be better for you to avoid drinking melting yellow snow as the presence of contaminants and harmful substances can be hazardous to your health.

Myth #5: Snow Is a Limitless Source of Clean Water

Reality: Snow cannot be considered an infinite water resource as, depending on your location, it may not even be available all year round. It should only be used in emergencies or when no clean, flowing water is available.

You need to educate yourself on safely consuming melted snow as a water source. Even in emergencies, take appropriate measures to ensure that the snow or ice you consume is clean. It is best to use the proper water purification method, such as boiling or water purification tablets, to treat the melted snow.

Potential Hazards of Consuming Snow

Sourcing water from melted snow and ice requires caution. It is essential for you to carefully melt the snow and purify it to avoid any health risks. Let's discuss the potential hazards of consuming melted ice water without proper treatment.

Hypothermia

You must beware of Hypothermia. Hypothermia refers to the state of your body when its core temperature drops below the normal temperature required for healthy metabolism and body functions, which is 95 degrees Fahrenheit (35C). All temperatures below 95F are considered hypothermic; however, it may vary:

- Mild = 90-95F

- Medium = 82-90F
- Serious = 68-82F
- Acute = 68F and below

The risk of hypothermia increases if you ingest snow directly, as it significantly reduces your body temperature. As snow's temperature is below freezing point, it can make your body expel extra energy to bring the temperature of snow down to normal body temperature. This leads the body's core temperature to enter hypothermic levels, easily worsening your bad situation. Hence, you should always try to melt the snow into water before drinking it.

You can make your water by putting the snow into a container or a bottle of water to make it melt and be good enough to consume. Please only use white, fluffy, and clean snow that does not seem dirty or contaminated. Purifying the melted water by boiling it before consumption is also a great idea.

May Contain Contaminants

It can be extremely dangerous for your health if you consume melted snow water without proper purification, as it may be contaminated. The water may contain pathogens like bacteria, viruses, and other chemicals. These harmful substances can get released into the snow from industrial and agricultural waste, animal waste, and air pollution. You should also know that the snow may also contain heavy metals, including mercury, lead, and arsenic. All these harmful substances can be released into the water when the snow melts, making it unhealthy to drink. The presence of these unhealthy substances in your water can be really bad for your health. This can cause many health issues as well, including neurological issues, gastrointestinal diseases, kidney problems, and even cancer. There may also be organic compounds in your water that can cause respiratory illnesses.

High levels of chlorine in your water can also be unsafe. If consumed in large quantities, it can lead to nausea and vomiting and may even cause liver damage. You can use a proper filtration system to remove bacteria, viruses, and other contaminants to make the water safe to drink. Boiling water even for one minute can make it safe for consumption.

May Lead to Dehydration

As discussed above in one of the myths of consuming melted snow water, drinking melted snow can lead to dehydration. As snow consists of a small amount of water, it can lead to quick loss of fluids and eventual dehydration. If you are going to consume snow, you must do it in moderation, as your body temperature may drop quickly if larger amounts are consumed at once.

Poor Sterilization

Properly sterilize the melted snow water for your safety's sake. As mentioned above, the snow may contain many contaminants, so you should try to purify the water properly through boiling or purification tablets. Poor sterilization may lead to the causation of waterborne diseases.

Melt snow carefully to avoid potential risks from consuming contaminated melted snow water. You must take sufficient precautions to ensure the water is safe for consumption. If possible, try boiling, filtering, or adequately melting the snow before drinking it. Taking these steps will significantly reduce the risks to your health.

Chapter 7: Conserving Water in Scarce Regions

This chapter focuses on strategies and practices to optimize water use in areas where resources are limited. After detailing the challenges unique to arid or drought-prone regions and the dire implications of water wastage, the chapter outlines actionable steps to minimize water usage, from daily activities to agriculture and industry. You'll learn about traditional water-saving practices, innovative solutions, emerging water-conserving technologies, and community-driven efforts and policies that have made a tangible difference in water-scarce regions worldwide in short and long-term scarcity situations.

Conserving water is essential.
https://pixabay.com/photos/to-protect-hands-ecology-protection-450596/

Water Scarcity and the Importance of Water Conservation

Due to a combination of extreme population growth, inefficient water management, pollution, and climate change, water scarcity has become a pressing issue in several areas of the world. Even if you don't live in a region of global water scarcity, contributing to conserving this highly valuable resource could mean a vast difference in the future. From hygiene issues to economic losses through unsustainable agriculture to the dire consequences for the Earth's ecosystem, water scarcity enormously impacts many areas of life. This problem is particularly notable in arid regions due to high evaporation rates and limited precipitation. Because of this, over 50% of the world's wetlands have already disappeared over the last century. Wetlands are the most sustainable habitats on the planet as they support the survival of many species. They're also the primary cultivation place for staple crops like rice and support a variety of ecosystems that benefit people, including flood control, water filtration, and storm protection.

In areas of water scarcity, all-natural landscapes suffer. Freshwater lakes are getting smaller and becoming salty due to pollution and because too much water (compared to the overall volume) is used for irrigation. Seas are retracting, leaving behind polluted land, creating food shortages and a rise in poor hygiene-related and waterborne diseases, and lowering the local population's life expectancy. Moreover, without adequate clean water resources, people can't get enough for consumption and agriculture, which leads to economic decline.

However, water conservation isn't only needed in drought-prone areas. Finding sustainable solutions to this issue is vital for improving water availability and conserving this resource. It also helps promote natural ecosystems, preserve habitats, and prevent saltwater intrusion into groundwater in coastal regions. Moreover, it can help save money for individuals and their communities and for entire countries by reducing the energy consumption needed for water transportation and treatment.

Practical Solutions for Water Conservation

Building Dams and Reservoirs

Establishing reservoirs like artificial lakes after river dams is an efficient way to collect water during high precipitation periods and conserve it for dry periods. However, they also have another purpose. They can be used as water supplies for urban areas, flood control, hydroelectricity creation, and more. Unfortunately, dams aren't the most ecological solution as they lead to downstream river channel erosion and other negative impact on the local ecosystem. The water in these reservoirs warms up more quickly, which means it evaporates rapidly and becomes unsuitable for species adapted to cold river water. Moreover, the sediment trapped in the damp creates a need for further filtration, depending on what the conserved water is intended for.

Reservoirs can help you collect and conserve water.
Naksh, CC0, via Wikimedia Commons: https://commons.wikimedia.org/wiki/File:Water-Reservoir-51552-pixahive.jpg

Desalination

Advanced desalination (removing salt and other compounds from seawater) technologies have become increasingly popular in coastal regions. Salt can be extracted from brackish and seawater through distillation or reverse osmosis, rendering it suitable for agricultural uses, like irrigation, human consumption, and industrial purposes. As saltwater sources are readily available in coastal regions, desalination represents a sustainable resolution to freshwater scarcity in these areas, reducing the dependence on fast-disappearing resources and helping meet the needs of coastal populations. Benefits notwithstanding, desalination has drawbacks, including brine disposal issues, which can affect the ecosystem and high energy consumption. Solar-powered desalination is a more cost-effective variant, as it uses less energy and can be applied in smaller regions.

Aqueducts

Installing aqueducts is another way to transfer water from one place to another. However, they only work in small areas because drawing water from larger areas could lead to drought in these areas in the future. These dry areas lead to the loss of ecosystems and significant air pollution in the surrounding region. Moreover, transferring from larger areas would mean crossing a vast distance, which raises the cost of the process.

Rainwater Harvesting

For centuries, people have been collecting rainwater for agriculture, recharge, and human use. In areas where rainfall is scarce, harvesting and storing it can be a fantastic solution for improving water availability during dry spells. The traditional method of rainwater harvesting involves collecting water from rooftops and similar surfaces that catch and lead large quantities of water toward the ground. This inexpensive approach can be used in off-grid situations as it can be combined with water purification techniques to make water safe for human consumption.

Beyond offering a secure water supply during droughts, the method also diminishes the reliance on other water resources. If an entire community starts implementing rainwater harvesting, they'll have a significant impact on improving local ecosystems in nearby wells, lakes, and rivers while contributing to long-term water management. Groundwater recharge, for example, enables water to move from the earth's surface to the deeper layer of the ground. Harvesting rainwater is a viable solution to create more groundwater and reduce water scarcity. Advanced rainwater harvesting methods go a step further. These also eliminate soil erosion, flooding, and other problems created by sudden stormwater runoff in urban environments and help use this excess water for building long-term storage instead.

Water-Smart Agriculture

Water-smart agriculture refers to targeted farming practices for boosting or retaining crop productivity while maintaining optical water usage efficiency. This approach involves both traditional agricultural methods and modern water conservation efforts like planting drought-resistant crops, drip irrigation, or precision agriculture.

Drip irrigation delivers water into the ground, directly to the crop's roots, maximizing water efficiency and minimizing losses due to evaporation. By enabling farmers to apply precise water delivery control and conserve resources, drip irrigation can lower water consumption by up to 50% while optimizing crop yields.

Xeriscaping is an agricultural method involving using crop fields requiring minimal water use and maintenance that uses water. These landscapes are created through a combination of drought-resistant plants, optimized mulching applications, and efficient irrigation systems to lessen water consumption. Beyond preserving water, xeriscaping also diminishes the need for pesticides and fertilizers, resulting in healthier and safer crops for human and animal consumption.

Adopting these and other water-smart agricultural practices can be an excellent way for farms to lower production costs, improve water retention in soils, contribute to sustainable water resource management, and reduce evaporation rates. These agroforestry practices have an overwhelmingly positive impact on ecosystem resilience and biodiversity preservation.

Pollution Control

Uncontrolled water pollution can lead to even greater water scarcity, especially in drought-prone areas. To prevent freshwater resources from becoming unsafe to drink and having to look for other solutions for an increasing number of people, sewage systems should be improved, and water quality should be regularly monitored and monitored.

Wastewater Treatment

Implementing efficient wastewater treatment solutions is another way to reuse and reclaim contaminated water. These techniques remove all the contaminants from water, producing a fresh supply of reusable water. In most cases, this water is used for industrial or agricultural purposes.

However, with the right technique, it can be purified to a level that becomes safe for human consumption. Moreover, modern water purifying techniques create a natural gas byproduct that can produce energy. By investing in this cutting-edge technology, people can contribute to a sustainable infrastructure for continually reusing resources and securing their water needs.

A form of water waste treatment is greywater recycling, which entails collecting used water from different sources in a household, including sinks, washing machines, and showers. This water is typically used for irrigation, groundwater restoration, industrial purposes, municipal water supply, and toilets, all of which reduce the dependence on freshwater supplies while reducing wastewater dump.

The best part of greywater recycling is that it can be implemented for both large-scale and small-scale users. For instance, in coastal areas, wastewater from a company's or municipality's waste system can be used for groundwater supplementation, limiting saltwater intrusion. Small-scale use is seen in single households, for example, where the plumbing is designed so that greywater from sinks is used to flush the toilet. For the most part, greywater is filtered until it becomes non-potable, which is unsafe for conception. However, if this treated water is sent through additional purification and aquifers, it can become potable (drinking) water.

Water Conservation

Water conservation entails using water more economically. For individual households, this means tailoring user behavior to use less water and implementing water-saving technology. The former can include washing dishes through a dishwasher instead of hand washing them, running the dishwasher and the washing machine only when these are full, so they require fewer washes which use less water than small frequent loads, taking short showers instead of baths, turning off the water while cleaning your teeth, or fixing leaky fixtures. Additional water conservation requests could include opting for produce with a reduced water footprint, like eggs, instead of high water footprint variants like beef. Likewise, growing drought-tolerant plants in your garden in dry climates, irrigating only as needed and preferably in the morning before the sun gets high and evaporation begins, can save surprisingly high volumes of water. Water-saving technology includes installing water-conserving shower heads and toilets and more efficient washers.

Since agriculture uses far more water than households, an even greater effort for water conservation can be made here. Besides the drip irrigation system, another viable solution could be growing only crops native to the area and only in places with enough precipitation to sustain them or using no-till methods, which rely on soil coverage application to reduce evaporation.

Smart water metering is an innovative water conservation technique that can be implemented in various fields. Industries, municipalities, and even individual households can monitor their water usage through real-time water consumption measurement. This enables users to pinpoint areas where they can curb or modify their water usage and management. By identifying excess water usage and implementing the necessary measures to reduce this and become more responsible, users can reduce their consumption by up to 20%.

Encouraging Water Conservation by Educating Others

While individual users can significantly contribute to water conservation, they can do much more to secure water supplies in the long run by educating those around them. By collaborating with others in your community, you can radically reform water consumption in the entire community. Educating others involves raising awareness about the importance of water conservation. They can also be encouraged to get involved in the process and provided with practical tips on saving water on their properties. Individuals should also aim to stay current with the innovative water conservation technology available in their area and how these can help resolve water scarcity issues. This fosters a sense of responsibility for their own behavior.

Educating community members about the impact of their water consumption habits and encouraging economical water use practices can be done in several ways, including media and social media outlets, school programs, etc. Community-wide awareness campaigns are another method for teaching individuals of all backgrounds about the importance of appreciating the value of limited freshwater resources.

The positive effects of collaborative efforts for this purpose have been proven time and time again, and not just on a community level. However, it all starts there. When communities raise an issue, other organizations and governments will notice it and start putting in the effort to exchange resources, expertise, and knowledge that brings the necessary changes to resolve the issue.

Past experience shows that governments that created and implemented enforceable legislation and policies to incentivize water-efficient practices have made a vast difference. Some of the best ways this was achieved was by fostering more responsible water management in industrial sectors and individual households and promoting the use of highly efficient technologies.

Not only can communities collaborate in water preservation efforts, but so can countries. Freshwater resources are often shared by at least two countries. Thus, it makes sense that they cooperate in managing this very finite resource. Promoting collaborative efforts like joint initiatives and water-sharing agreements could foster a greater sense of collective responsibility.

Taking Advantage of Artificial Intelligence

Artificial Intelligence (AI) has revolutionized many sectors, including water management. Utilizing AI to supplement water conservation efforts helps people to make more informed decisions on researching and developing new water-saving technologies and techniques. AI can help analyze water consumption data and optimize water management strategies based on predicted water demand, weather forecasts, insights on optimal irrigation schedules in agriculture, possible issues (like leaks, for example) in building infrastructure, and more.

By leveraging these tools, entities can make better decisions regarding conservation efforts, investment in improving infrastructure, and water allocation, addressing large-scale water scarcity challenges and securing sustainable water use.

Real-Life Examples from Around the World

Langtang Region, Nepal

As remarkable as the landscape in the Eastern Himalayan regions is, the rugged terrain and climate changes have caused many headaches for local farmers. The weather shifts from extreme droughts through the winter to prolonged monsoon seasons through the warmer months. It makes it harder and harder for the locals to grow crops and sustain their livelihood and the traditions that have depended on the area's natural resources for centuries. On top of that, the mountain glacier lakes are melting, threatening floods and the reduction of freshwater resources that currently sustain over a billion people. It could additionally destroy the habitat of numerous species already endangered by human actions.

To address the issues mentioned above in their area, communities in Nepal's Langtang region, along with the support of outside organizations, developed a project to help them better adapt to climate changes, including unpredictable precipitation patterns, water scarcity, and shifting seasons. These changes included installing drip irrigation systems (extended across multiple communities), along with water storage to preserve rainwater, establishing seed banks to supply farmers with more drought-tolerant species, and education programs to teach them how to implement techniques like crop rotation and reduce water footprint when raising farm animals.

Additionally, the efforts included planting shrubs and trees along slopes to prevent landslides and endangering freshwater resources. The villagers were given more efficient household appliances, and newly established weather stations gathered data to monitor subsequent climate change. All these efforts helped educate the community, making a vast difference in the villagers' lives and ensuring their livelihood and the livelihood of their families. The project did not only address the effects of climate change but also the needs of people and the local ecosystem. It raised people's awareness and gave them a sense of responsibility for preparing for the inevitable future effects of climate change. It also served as an example to other Himalayan communities facing the same challenges, which are now working on implementing similar changes.

Sahel Region, Africa

The Sahel region of Africa faces a similar problem, receiving rain only three months of the year. However, while this wasn't usual in the past, the periods of rain have become unpredictable due to accelerated climate changes. During the months without precipitation, the soil dries out completely, becoming so hard that it can't even penetrate the surface layers when the rain arrives. Instead of giving it much-needed hydration, the water simply slides off the surface and evaporates under the scorching sun. Consequently, local communities struggle to secure water for the better part of the year for agriculture and hygiene, cooking, and drinking. Fortunately, they received help from several organizations and learned how to conserve water during those short months of the rainy season so they could have water to use throughout the rest of the year. For example, they've learned to build pits (zai or half-moon pits) – essentially holes dug into the soil to trap rainwater and prevent it from sliding off the hardened surface. Stone binding barriers are another example of the effort to keep the water on the agricultural lands.

Trees around lakes are planted to stabilize the land and prevent it from being washed away during heavy rainfalls. Several communities are working on digging even larger water storage pits. These 10-foot-deep and 100-foot-wide reservoirs represent more targeted efforts as they will be established after a thorough evaluation of where water is collected during heavy rainfalls, calculated based on the terrain and the water's natural flow. These tanks will be able to hold the same amount of water that could fill an Olympic swimming pool, providing a substantial resource for families struggling to meet their freshwater needs throughout the year. Due to their size, the water in these reservoirs won't be affected by evaporation despite the extremely high temperatures and the unpredictable rainfall.

The benefits of the large rainwater pits in the Saharan regions will be short- and long-term. By putting water conservation techniques in place, the local communities can secure access to fresh water all year round, contributing to public health and safety and ensuring the productivity of the lands that sustain their livelihood. Moreover, the soil quality around the reservoirs improves as the rain gets trapped in the deeper layers. People can start growing crops in previously dried-out lands and obtain more nutritious food resources. This effort can be further bolstered by planting an outer ring of trees around the reservoirs. Trees also trap water and provide shelter for the crops, preventing the water used for their irrigation from evaporating. This shows that water conservation is not only beneficial for having a resource for a community to tap into but can also transform the entire environment and create new, more sustainable ecosystems.

Chapter 8: Surviving While on the Move

Detailing the unique challenges and strategies associated with maintaining hydration and sourcing water when continuously mobile, this chapter provides a better understanding of the preventive and reactive measures to stay hydrated while moving through diverse terrains and situations. From water rationing to purification to reading natural signs that can hint at nearby water sources, you find everything you need to ensure long-term water sustainability without guaranteed resupply.

The Importance of Staying Hydrated While Being Mobile

Whether you are forced to be on the move for migration, long-distance trekking, or fleeing emergencies, staying hydrated is vital, especially outdoors. Knowing how to avoid dehydration is key to enjoying outdoor activities safely and staying healthy in emergencies. Water is needed for many metabolic and physiological processes in the body, many of which are critical for survival. Depleting your body's water storage results in fatigue, loss of focus, heat exhaustion, and much more. For all these reasons, staying hydrated is critical for preparedness in extreme conditions or crises.

Staying hydrated is critical for preparedness in extreme conditions or crises.
https://pixabay.com/photos/faucet-fountain-water-1684902/

At the same time, the wilderness and other emergency situations make it far more challenging to maintain appropriate hydration levels. However, with proper preparation and by learning what precautions to take when facing extreme conditions and limited water resources, you can avoid dehydration and its consequences.

Tips for Staying Hydrated on Outdoor Adventures

Drink before You Start Moving

If you're going on an outdoor adventure, like hiking or a short, day-long camping trip, drink one or two cups of water, even if you aren't thirsty yet. When you start feeling thirsty, your body's water levels are already depleted enough for dehydration to begin. Drinking enough water before you go out will keep your body's water levels in an optimal range for longer. If you regularly hydrate before starting your day or activity, this will improve your fitness levels, enabling you to focus better while outdoors.

Suppose you're heading out in the morning. In that case, you can start hydration the evening before by ensuring you won't go to bed thirsty and avoiding caffeine and alcohol at nighttime. The latter two cause dehydration, so these are not great options to drink during camping or other outdoor adventures.

Make Your Portable Water Supply Easily Accessible

Make sure you carry ample water for your adventure. Besides this, you'll also need food, as this is a source of electrolytes and essential salts – which are just as crucial for preventing dehydration as drinking enough water. Sweating causes your body to lose electrolytes and leads to an imbalance in your hydration levels. For a short hike, a bottle of water and a few energy bars or other snacks or dried food will suffice, while a camping trip involving mealtimes requires food and water for each mealtime. If you're going on an extended adventure, count on meals and water for every day, plus a little more in case something comes up and your return is delayed. Electrolyte tablets are another option for these situations, and they can come in handy for fast rehydration in a crisis where food isn't available.

To ensure you'll be drinking and eating regularly during your adventure, make your resources easily accessible. While many prefer to carry water bottles, these make it harder to drink on the go. Bladders, however, can be easily used while on the move, ensuring you won't forget to use them. If you still prefer bottles, put them where you can easily reach them. For example, many hikers and campers prefer forward packs with forward-facing pockets instead of side pockets for water bottles because they make reaching for water easier. Using water-storing containers with built-in handles and attached carabiners is another alternative. You can clip them onto your belt or backpack and make them easily accessible while on the move. Likewise, keeping snacks in easily accessible pockets will make it more likely for you to reach them.

Drink Water Regularly

Instead of drinking when you feel thirsty, drink ½ -1 quart of water every hour while outdoors. Suppose the temperatures are high, and you're moving through demanding terrain. In that case, you'll likely need more because moving in warmer climates raises your body's water intake needs. If you're bored drinking water this frequently, you can alternate between pure water and electrolyte-filled fluids, like sports drinks. This will help you retain optimal hydration levels and make your outdoor adventure more enjoyable.

Stay Hydrated Afterwards, Too

Continue replenishing your water and electrolyte levels even after you return home, and don't wait until you get thirsty. Remember, thirst isn't a good indicator of your body's proper water needs, so it's always better to drink a little more rather than not drinking enough.

Signs of Dehydration

Dehydration should be treated as soon as you feel it because the longer you wait, the greater the consequences. Here are some signs of dehydration to look out for in yourself and those who are with you in your outdoor adventures or emergency situations:

- **Dry Mouth**: Before your body even registers thirst, you'll probably notice you have a dry mouth.
- **Thirst:** Feeling it means your body is telling you that you are dehydrated and should drink immediately because your water levels are about to become dangerously low.
- **Low Energy Levels:** You may assume you're tired because you're on the move a lot, but this can also be a sign of dehydration.
- **Muscle cramps, Headaches, and increased fatigue** are severe signs of dehydration. If you feel these symptoms, you must rest and rehydrate immediately.
- **Dizziness, Nausea, Stumbling, and Fumbling** are likewise serious symptoms, indicating that your body is struggling to provide enough fluids for normal metabolic and physiological function.
- **Dark Urine and Abdominal Pain:** Your kidneys suffer from a lack of hydration, making the urine concentrated and struggling to filter metabolites properly.
- **Weight Loss**: Even if you can't weigh yourself regularly, if you notice that your clothes fit looser after strenuous activity, it means you've lost water weight, and this water needs to be replenished through fluids and electrolytes.

What to Do in Case of Dehydration

If you notice the signs of dehydration, stop and find shade (if this is possible). Take a few moments to rest, especially if you have moderate signs like fatigue or muscle cramps. If you have more severe symptoms, you'll need to rest longer, and if the temperature is extremely warm, apply body cooling methods as well. Soaking a shirt or any other piece of clothing in water and placing it on the back of your neck, wrists, or head are the quickest ways to cool down. Meanwhile, drink water mixed with electrolyte tablets (if nothing else is available, table salt will do as well). Oral rehydration and salts help the body absorb water more effectively. Adding them to your first aid kit (a highly recommended part of your gear for wilderness adventures) won't take much space, and they can be a lifesaver.

A compact quick water purifier is valuable as it can make an enormous difference in your ability to rehydrate yourself or anyone else in case a fresh water supply isn't available. While you should always aim to have enough water, sometimes it can be difficult to judge how much water and electrolytes you need depending on the weather conditions, not to mention the possibility of emergencies in which you may need to use water for other purposes. Having a purifier handy will make rehydrating from alternative water resources easier without exposing you to pathogens and toxic substances.

Additional hydration tips for every outdoor situation:

- If you have plenty of freshwater sources on your route, drink about 33 ounces of water every time you stop to refill. This way, you'll remain hydrated while keeping your load light.

- Drink regardless of the weather. Unless you are forced to ration your supply, proper hydration is just as critical in cold weather as it is when temperatures are high. Not only does cold cause dehydration, but being dehydrated will make you feel like you're freezing. Use a thermos to keep yourself hydrated and prevent your water from getting chilled during the winter.

- Exposure to the sun will also aid dehydration, so wearing sunscreen and sun-protective clothing is essential for summer outdoor adventures. Even shade from your clothes and hat will help you regulate your body temperature, sweat less, and remain hydrated.

- It's a good idea to go on outdoor adventures in extremely hot climates when it's cooler. If you're on the move due to an emergency situation, try keeping to the shade when the temperatures are the highest and walking around in the morning or early evening when the sun is no longer at its peak.

• Rehydrate with at least 16 ounces of fluids (ideally mixed with electrolytes) overnights or when resting midday during the warmest hours.

What to Do if You Run Out of Water

If you notice that your water supplies have diminished or are about to run out, don't panic. Remaining calm will help you maintain your focus on finding alternative resources. Panicking is a waste of energy and time. It also increases your body's evaporation rate, making you lose liquids you can't afford to relinquish. As a first step, assess your situation and area. If you've just run out of water and it's not extremely hot outside, you'll likely be fine without water for a few hours. It may get uncomfortable, but if you've hydrated properly until then, it shouldn't be dangerous for your health. However, if you're further than that from the nearest water source or battling scorchingly high temperatures, you'll need to start searching for alternative solutions immediately.

Finding and Using Water in the Wilderness

Access to water is crucial for surviving in the wilderness. You'll need to find alternative sources when your supply becomes limited. Here are a few tricks on how to do that.

Look for Vegetation

Fruit can be a great water source - as long as you know which to look and reach for, as not all are safe for consumption. Familiarize yourself with the native flora in the area you'll be moving through during your outdoor adventure or mobile survival situation to learn which fruit and plants to look for as a water and electrolyte resource. Due to their high water and essential nutrient content, citrus fruit, grapes, berries, apples, and melons are the most recommended options - but there are many more you can reach for rehydration, including edible grass and moss variants. To determine whether a plant is suitable for this purpose, examine it for signs of water storage. For example, some plants store water in their nodes (where the leaves grow from. These are typically swollen when they're full of water), in the leaves (look for thick, moisture-laden foliage), or underground bulbs and tubers. The vegetation's color, shape, and size also play a role in determining its suitability for safe rehydration. For instance, brightly colored plants are often toxic, so you should avoid them. Moreover, some plant parts require preparation for either extracting water or rendering them usable for consumption.

Vegetation can lead you to a water source.
https://unsplash.com/photos/green-plants-and-trees-during-daytime-GwDOjh66Le8?utm_content=creditShareLink&utm_medium=referral&utm_source=unsplash

Tapping into trees is another efficient strategy for finding drinking water in the wilderness. Trees with edible sap, like maple and birch, are great options as they contain larger quantities of readily drinkable fluids. They are also easier to tap, and extracting the sap takes only a few minutes. For example, other trees, like walnuts, are also viable options, but extracting sap from them is far more challenging.

In arid areas, any sign of vegetation indicates moisture. Looking for livestock or other animals is another option, as these could lead you to a nearby water source (but be cautious when approaching or following them).

Find Flowing Streams

If you have rivers or other flowing streams in the area, these can be an even more reliable source of drinking water. Because water travels at higher speeds, flowing water sources contain less bacteria than standing water reservoirs. However, you still must be cautious when deciding which stream to use. For example, if you find evidence of animal or human waste near the source, or the water isn't particularly clear (fresh, unpolluted flowing streams are always so clear you can see through them), you should probably stay away from it - unless you have no other option and have the ability to purify the water.

An additional tip: If you find a small stream that's dried up, tracing its path upwards to its source can help you find an additional stream you can use for hydration. If you can find current water sources, you can try digging for water in moderately moist areas near dried-out streambeds.

Collect Rain and Snow

Rainwater can be a priceless resource for drinking and hygiene in unexpected situations during outdoor adventures and emergencies. To collect rainwater, set out large containers like buckets or barrels to store it until needed. You can also gather rainwater on the top of your tarps (or if you don't have any containers, you can hang strategically placed waterproof materials instead for collecting precipitation). Although rainwater is relatively clean, it should be further purified to eliminate any bacteria or other pollutants it may contain. The same applies to snow, another alternative water source, especially in mountainous regions. Snow and ice can be used only, if not too gray or yellow, and should be melted before consumption, as the cold fluid can lead to dehydration.

How to Purify Water

If you don't have access to clean water in the wild, your next best option is to purify the available resources, making them safe to drink, cook, or prepare medicine if needed. Depending on your specific circumstances, you can opt for several purification techniques, including the ones listed below:

- **Boiling:** The simplest and often most effective way to purify water. You'll only need to strain it through a small hole sieve to remove bigger chunks of impurities and bring it to a rolling boil for several minutes to kill all the bacteria and other hazardous microorganisms.

- **Activated Charcoal Filters:** Moving the water through activated charcoal filters will bind the harmful organic compounds and remove unpleasant odors and tastes from the liquid.

- **Purification Tablets:** Given that these tablets can clean large amounts of water in a short period, they are an excellent addition to any outdoor survival gear. Place one tablet into half a gallon (or a little less if the water is particularly contaminated), wait 30 minutes for the tablet to neutralize the harmful compounds, and your water will be ready for consumption.

- **UV Purification:** UV purifiers emit bacteria-killing radiation that can eliminate these microorganisms from your water supply in minutes.

Making the Most of Your Water Supply

Carrying Enough Water with You

Carrying enough water is one of the most fundamental factors to consider when preparing for an outdoor adventure or emergency survival situation. First and foremost, you must assess how much water you'll need for the time frame you'll be out on the move. This depends on several aspects of your situation, including the required physical activity levels, climate, the difficulty of the terrain, altitude, whether you'll be traveling during the day or night, whether you can expect obstacles in finding reliable food and water sources, and many more. For example, a person with a moderate activity level hiking on a demanding terrain on a warm spring day will need 33 ounces of water for every 2 hours of activity.

Refill Whenever Possible

Filtration kits, gear, and bottles make it easy to replenish your water resources when you walk past or near a freshwater resource in the wilderness. Look for collapsible options, as these can be stored in your bag when empty and used when needed.

Before heading out, research possible water sources nearby, but in the wilderness and from human origins. For example, you might find houses or camping grounds with water outlets nearby or temporary abandoned dwellings with running water or stored water resources. Have a plan in case anything goes wrong in an emergency or survival situation. Knowing where to go for a refill makes it easier to remain calm and preserve your energy and fluids.

Rationing Your Supplies

Suppose you are still worried about running out of fresh drinking water or don't have time to devise a refilling plan (or to research the area for available resources). In that case, you may need to ration your supply until you develop an alternative solution.

At moderate activity levels and temperatures, drinking only when you're thirsty is one of the best ways to preserve water. Rationing your loss rather than your supply will help you stay hydrated, waste less, and worry less about your dwindling resources. Covering up in hot and cold water will lower the risk of dehydration, and so will remaining in the shade. Your supply will last longer with fewer losses due to the sun, wind, and other elements.

Another tip to minimize evaporation from your body is to keep your mouth closed whenever possible. Unlike breathing through the mouth, breathing through your nose results in a slower and smaller evaporation rate. If you struggle to keep your mouth closed, try pretending that you're sucking on a round candy. This will automatically maintain your mouth in a closed position. Sucking also fosters increased saliva production, which will prevent your mouth from feeling dry, making you crave water less often. However, avoid actually sucking on pebbles or buttons because these can be a choking hazard if you are severely dehydrated and unfocused.

Smoking also adds to dehydration through evaporation, so you should avoid it if possible. Coffee, teas, and other drinks with a diuretic effect increase the amount of urine produced by the kidneys, leading to significant water loss. The same applies to salty food and drinks because these draw water away from cells.

Another crucial tip is to start saving water in time. Don't wait until you are severely dehydrated to ration your supply! By implementing water-saving methods from the get-go, you can save more efficiently and avoid getting into potentially dangerous situations.

Chapter 9: Long-Term Water Storage and Safety

In survival situations, whether it's dealing with a natural disaster, a prolonged emergency, or an outdoor adventure, having a plan for keeping a stable water supply is key to your well-being. This chapter will cover the ins and outs of long-term water storage. You'll learn about the methods for doing so and the important things to consider when it comes to maintaining a reliable water source for an extended period. How do you make sure the water you store remains safe to drink for weeks, months, or even years? What factors affect how long your stored water stays good, and how do you reduce the risk of it getting contaminated or going bad?

From understanding water quality to choosing the right storage containers and managing environmental factors, this chapter will prepare you to face the challenges of keeping a sustainable water supply, blending scientific know-how with real-world advice. Remember that in the world of survival, thinking ahead for the long term is just as crucial as having enough water.

Keeping a stable water supply is key to your well-being.
https://pixabay.com/photos/storage-tanks-vats-metal-tanks-20959/

Water Preservation

In the previous chapters, you learned the difference between disinfection and sterilization when it comes to water treatment. The goal is to make raw water safe for consumption by either killing or removing harmful components. However, it's crucial to understand that even the methods outlined in this book, including commercially available ones, don't produce sterile water completely free of germs or viable germ stages.

Even if you were to achieve sterile water, it would quickly become recontaminated when transferred to a different container, which could have stray dried-up algae, protozoa, or spores from the air in it. The methods described previously aim to purify or reduce the germ count in raw water, making it safe for consumption. With improvised methods, you can realistically reduce the germ count by a factor ranging from a hundred to ten thousand. Even sterilization in a laboratory rarely eliminates germs entirely, usually resulting in a reduction by a factor of one million at most, even under ideal conditions.

Now, let's talk about recontamination. The drinkability of water treated with improvised methods is not a given and depends on the individual's health and conditions. After the sanitizing process, the remaining germs continue to multiply, and their growth follows an exponential pattern. The speed at which this happens depends on factors like nutrient levels, temperature, and the initial germ count.

While there's no immediate need to drink or discard the treated water after a few hours, it's essential to consider that the remaining viable germs will continue to multiply. However, the risk can be minimized by storing treated water in clear conditions, carbon-filtering it if possible, and keeping it in a shaded or cool place. This generally ensures it remains safe to drink for weeks without harmful pathogen levels. This is why the storage of water is a process not to be taken lightly.

One remaining concern is bacteria that can tolerate normal temperatures and have modest nutritional requirements, like Legionella. However, the risk of harmful levels is significantly reduced with proper storage conditions and limited nutrients. Below are some of the most effective ways to preserve water and limit its contamination:

1. Tyndallization

Tyndallization is a laboratory process used to disinfect sensitive solutions. It involves repeatedly heating the solution to boiling point and then holding it at body temperature (98.6°F or 37°C) for a period called *incubation*. Boiling eliminates all relevant living germs but doesn't remove nutrients from the water. The heat triggers inactive endospores to "germinate" upon cooling, transforming them from a resistant dormant stage into a less resilient active one. They are then killed when the water is boiled again. This cycle is repeated multiple times. In a modified form, this method is used, considering that laboratory solutions, like nutrient broths, have a higher risk of rapid recontamination than relatively nutrient-poor drinking water.

Water heated to boiling can be consumed immediately after cooling, but it quickly recontaminates as nutrients from destroyed cells become available. Briefly reheating the water over the next two or three days reduces the number of inactive and heat-resistant stages to very low levels. After Tyndallization, the water can be stored and consumed at room temperature for several weeks.

2. Repeated SODIS

After the correct application of Solar Disinfection (SODIS) for a sufficient period, water remains free from harmful germs for several days. UV light, the primary agent in SODIS, can even kill stable endospores and oocysts of cryptosporidia and giardia. However, if the raw water is still cloudy or if germs are hidden, recontamination with bacteria is a possibility.

Exposing water stored in suitable containers to direct sunlight for an hour or two at midday kills any new or remaining germs or endospores. After repeating this process two or three times, the water can be transported without refrigeration for several days or weeks, depending on its nutritional content. This method is especially useful for prearranged desert trips, where clear PET bottles filled with reservoir water can be stored directly in the sun. However, prolonged exposure to UV radiation can age the plastic material, so these stores should not be left for more than a year. The water inside remains drinkable.

3. Chilling and Shading

Water treated with an "unreliable" germ count reduction method is likely to still contain algae, leading to the growth of fresh biomass in the presence of sunlight, nutrients, and dissolved gases. This results in a musty, putrid flavor and a slimy, foul-smelling layer (biofilm) forming inside the container. While this water is not immediately unfit for consumption, it becomes a breeding ground for any remaining viable, infectious germs over time, especially at temperatures above 59°F (15°C).

To counteract algal and bacterial growth, storing the water in a cool and dark place is crucial. This can be done by covering it with wet sand, which chills the container through evaporation, or by wrapping it in a rescue blanket and placing it in cold, raw water. Depending on the initial contamination level, water treated with "unreliable" methods can be stored for around two to three days.

4. Silver Ions and Other Chemical Substances

Similar to disinfectants, there are chemicals available for preserving drinking water. However, these agents are not considered reliable disinfectants. While disinfectants with oxidizing substances can kill and destroy pathogens, they cannot prevent the regrowth of any potentially remaining germs permanently. The reactive properties of oxidizing substances mean they do not remain stable for long once activated.

Water Storage Techniques and Ideas

1. Emergency Water Pouches

In today's world, convenience is everything; you can find practically anything prepacked and ready to go. Even drinking water comes in individual ready-to-drink portions—small pouches or packets holding about 4.22 ounces (125 ml) of water. However, these can be relatively pricey, ranging from 3 - 10 US dollars per liter. While marketed for their long shelf life and touted as emergency reserves for your car or boat, there's a more cost-effective and equally stable alternative: regular small water bottles, usually 330 or 500 milliliters (11.2 or 16.9 fl. oz.), found in any supermarket worldwide.

The water in these bottles is actually sterile, evidenced by its extended shelf life. You can safely ignore any use-by date printed on them—water safe to store for months remains just as safe to drink after decades. The only changes you might notice are a slight plastic taste and a slow reduction in quantity due to diffusion.

In contrast, "normal" flavored or sugary drinks have a shelf life. Over time, decomposition sets in, rendering the contents unfit for consumption. Carbonated water loses its fizz gradually, as the bottles aren't airtight, but other than a taste change, the quality remains relatively unchanged.

Here's the scoop on emergency water pouches: they're essentially tap water sealed in shrink-wrapped pouches. The good news is you can easily create your own at home. Just vacuum-seal ice cubes in a plastic bag, let them defrost, and "autoclave" them for half an hour in a pressure cooker. Consider storing fruit or iced tea mix powder in sealed jam jars or pouches if you like the idea of having flavored or sugary liquids on hand for emergencies. Add a bit to your water when needed, and you're good to go. It's a budget-friendly and practical way to ensure you have emergency hydration options without breaking the bank.

2. Transport and Storage Containers

In emergencies, ensuring a reliable and substantial water supply is critical. While water is often a transient necessity, certain situations demand preparedness. Stationary storage tanks are a valuable option for stand-alone systems in remote camps, disaster areas, or buildings without water supply connections. These large tanks, often intended for temporary use but capable of permanent installation, can be filled naturally using rainwater-harvesting systems. Due to extended standing periods, the water they hold must be treated as raw water rather than fresh rainwater, allowing for the storage of significant amounts over months or even years.

In the past, rural areas and off-grid regions stored water in cisterns, but today, plastic tanks are more prevalent. Both have pros and cons. Building a stone or concrete cistern presents challenges, while plastic tanks, such as IBC tanks with a capacity of around one cubic meter (250 gallons), are more susceptible to frost damage.

- **Cisterns and Ponds:** Ideal for holding large volumes of water, but even initially, clean water must be treated as raw due to its open nature.

- **IBC Tanks:** A practical choice for stand-alone systems. Sealable and stackable but vulnerable to freezing and bursting in low temperatures.

- **Buried Plastic Tanks**: A mid-sized solution that can be buried below the frost line. Less prone to bursting if it freezes, they can be filled with rainwater or covered with a roof to prevent contamination.

3. Mobile Medium-Size Storage Containers

Mobile storage containers, commonly known as *canisters*, become indispensable for those traversing arid regions by car or cart. With a carrying capacity ranging from five to twenty-five liters (one to six gallons), these containers are versatile and valuable for storing locally collected or harvested water in water-scarce regions.

Two primary options for transporting water are rigid plastic or metal canisters and folding water carriers. It's advisable to steer clear of collapsible water containers popular among campers because their thin walls are prone to breakage along folding lines during prolonged use.

• Rigid Water Containers

Rigid containers, while sturdy and easy to carry, come with a caveat—they occupy the same space, whether full or empty. Stackable and relatively robust, they often feature an integral tap that may flip open or snap off. Some prefer new "jerrycans," typically used for fuel, to avoid residual fuel contamination risk. Ensure food-safe canisters with quality lid seals are chosen to prevent leaks and maintain water quality.

Plastic containers have drawbacks, as they can absorb tastes and odors from adjacent canisters. Clear labeling and segregation (fuel/raw water/drinking water) are essential, and water containers should be stored away from fuel and lubricants.

• Water Carriers

While more susceptible to damage than canisters, water carriers have a distinct advantage; they only take up as much space as the water they hold. When empty, they can be conveniently rolled up and transported. Two common types include the Swiss Army water carrier (20 liters or

5.28 gallons), made from robust rubber with a solid spout, and water bags with a large filler cap, crafted from thinner material and available in various sizes.

- Swiss Army water carriers, while reasonably priced, can heat up in the sun, affecting water taste.
- Water bags made from synthetic material, like those from Ortlieb (a German manufacturer of outdoor equipment), are lighter but more expensive and prone to damage.

Bottles and Drinking Vessels

- **Transparent PET Bottles**: Lightweight, affordable, and robust, they're excellent for simple drinking and storage. However, they're heat-sensitive and unsuitable for boiling water.
- **Wide-mouthed polycarbonate (PC) Bottles:** Ideal for hot water and boiling without deformation.
- **Metal Bottles:** Allow for controlled water consumption and can be heated in a campfire. Suitable for serious physical separation and treatment methods.

Metal bottles allow for controlled water consumption.
https://unsplash.com/photos/green-hydro-flask-tumbler-on-wood-slab-ktpymCAIvGc?utm_content=creditShareLink&utm_medium=referral&utm_source=unsplash

Choosing the right bottle depends on water availability and expected treatment methods. Transparent bottles are preferred in water-deficient areas, while metal bottles are suitable for physical water treatment methods. Always consider the specific demands of your journey and the type of water treatment you plan to apply.

Improvised Transportation and Storage Methods

For backpackers with limited space and transport options, especially those on long journeys, water bags become a handy choice for collecting raw water or carrying emergency drinking water. But there's a catch – water bags have their limits, especially if unexpected rainfall offers a chance to store more significant amounts. Here's where some bushcraft experience can come in handy: you craft containers from bark, reed, bamboo, or wood. Remember, though, that natural materials can swell when wet and crack when dry. That's where thin, lightweight plastic sheets come to the rescue, serving as a reliable travel companion for extreme and long-distance travelers, facilitating the safe transportation of water.

1. Crafting Storage Ponds with Simple Tools

For those seeking an improvised storage solution for a temporary camp, a rescue blanket or similar sheet can be transformed into a small "pond." Start digging a shallow pit, ensuring the ground is free of sticks and sharp stones. Compact the soil thoroughly to prevent it from giving way under the water pressure. The pit's dimensions should allow the sheet to extend beyond the edges, giving you storage of fifty to a hundred liters (fifteen to twenty-five gallons) of water. These improvised vessels, created with natural materials, can be swiftly fashioned with lids, combined into transportable batteries, and even repurposed for cooking.

2. Emergency Containers for On-the-Go

If space in your backpack is scarce, choosing a lightweight water bag is not feasible. In such cases, you can resort to purchasing cheap wine bladders, often found in wine boxes. These are made from thin plastic but are prone to damage. Alternatively, large freezer bags and condoms, often part of emergency kits, can be used. However, these containers risk bursting even with minimal impact. Make a simple cloth bag from a T-shirt or unzipped trouser leg to stabilize them. Tie one end of the cloth in a tight knot, place the empty waterproof material inside, and fold the top seam to protect it. Fill the improvised carrier with water, ensuring no air pockets remain, and secure it with a knot or string. Lift the outer bag, tie it to the inner bag, and voilà – a stable, improvised water carrier ready for transportation. It's advisable to carry it outside the backpack or in your hands to avoid accidental spills and make the most of your resourceful water-carrying solution.

Additional Tips for Water Storage

Water storage is a critical aspect of preparedness, and understanding the nuances of storage practices in diverse environments ensures a reliable and safe water supply. Factors such as temperature, light exposure, and the threat of pests play pivotal roles in determining the efficacy of storage methods. Below is a comprehensive guide outlining best practices for water storage in different environments:

1. **Temperature Considerations**
   - **Cold Environment**
     - **Insulated Containers:** In colder climates, insulation is key. Use double-walled or insulated containers to prevent freezing.
     - **Underground Storage:** Burying water containers underground helps maintain a more stable temperature, preventing freezing.
   - **Hot Environments**
     - **Dark Containers:** Opt for dark-colored or opaque containers to minimize light exposure and algae growth.
     - **Shaded Storage:** Keep water containers in shaded areas to reduce temperature fluctuations and evaporation.
2. **Light Exposure**
   - **UV Exposure**
     - **Dark Containers:** UV rays can degrade water quality. Choose dark-colored or opaque containers to shield against sunlight.
     - **UV-Resistant Materials:** Invest in containers made from UV-resistant materials to ensure long-term water quality.

- **Indoor Storage**
  - o **Opaque Containers**: Even indoors, use opaque containers or store water in a cool, dark place to prevent light-induced degradation.

3. **Pest Prevention**
   - **Insect and Rodent Control**
     - o **Sealed Containers**: Use tightly sealed containers to prevent insects and rodents from contaminating the water.
     - o **Elevated Storage:** Elevate containers to deter rodents and use traps or repellents in storage areas.
   - **Filtering**
     - o **Mesh Screens:** Place mesh screens or filters over openings to prevent insects and debris from falling into your container.
     - o **Chemical Treatments:** Consider using safe chemical treatments that deter pests without compromising water safety.

4. **Container Selection**
   - **Material Considerations**
     - o **Food-Grade Containers**: Always use food-grade containers to avoid leaching harmful chemicals into the water.
     - o **Plastic vs. Metal**: Plastic is lightweight and convenient but may degrade over time. Metal containers are durable but may impart a metallic taste.
   - **Rotation and Inspection**
     - o **Regular Rotation:** Rotate stored water regularly to ensure freshness and prevent stagnation.
     - o **Inspect for Leaks:** Regularly check containers for leaks or damage that could compromise water quality.

When dealing with water storage, make sure you always consider your environment. If you're dealing with extreme temperatures, choose containers that can handle the heat or provide insulation against the cold. Opt for materials that maintain the quality of your water, and conduct regular checks to ensure everything stays in tip-top shape. Vigilance is key. Regularly inspect and rotate your water storage to prevent any surprises. Prioritize containers made from food-grade materials to ensure water purity. Think of your water supply as an investment in resilience rather than just a backup plan. Consider long-term strategies. Think about preservatives for extended storage, and explore crafting emergency water pouches at home for added preparedness. Functionality matters – choose containers with reliable taps, especially in accident-prone areas.

Chapter 10: Beyond the Bottle: The Many Uses for Water

Beyond its foundational role in hydration, water has versatile uses. From making personal hygiene more accessible to facilitating industrial processes, having access to water is a must. There are numerous examples where you'll learn that carrying out these processes becomes virtually impossible without water. Although humans have relied on water for agriculture and household needs for centuries, most industries require water, including food processing, chemical production, smelting, paper making, commerce, and much more.

Water in everyday life.
https://pixabay.com/photos/water-jet-shower-to-water-water-8007873/

The chapter digs into the practical aspects of leveraging water for survival, shedding light on its multiple dimensions as a crucial resource in the face of diverse challenges. This guide illuminates the role of water in fostering hygiene, emphasizing its significance in preserving health and well-being, especially when conventional medical resources may be scarce.

Water in Everyday Life

Hygiene and Personal Care

As you explore the many uses of water, hygiene stands out as a basic aspect of daily routines. It is essential for bathing, washing hands, and keeping the body clean. With adequate personal hygiene, harmful bacteria and microorganisms get washed away, protecting the body from diseases and further spread. If you're not maintaining hygiene, your body becomes susceptible to respiratory infections like colds and flu, diseases of the alimentary system may emerge, and even increases the chances of developing viral infections.

Cooking and Food Preparation

Besides its role as a simple ingredient, water becomes an essential component in the preparation and rehydration of food. While in survival situations and some cultures, cooking meals without water is a norm, most regions count water as an essential ingredient, making food palatable, nourishing, and easily digestible. Water is necessary for these cooking methods, whether boiling, simmering, or steaming. Most consumables can't be cooked properly without water or may lose nutritional value.

Cleaning and Sanitation

From cleaning the household to washing kitchen utensils, water is a powerful and natural ingredient that can be used for all your cleaning and sanitation needs. It is a universal solvent that removes dirt, grease, and other impurities, making your indoors squeaky clean.

Agriculture and Irrigation

It's the lifeblood of crops in agricultural practices. Water creates the correct soil moisture to promote seed sprouting root formation and supply essential nutrients. Water is also soaked up by the roots of already developed crops, promoting their healthy development and better yield. Plants can't grow, thrive, and produce an adequate yield without water.

It is impossible for plants to grow without water.
Photo by Tony Pham on Unsplash https://unsplash.com/photos/woman-in-black-long-sleeve-shirt-and-black-pants-standing-on-green-grass-field-during-daytime-TV7m_tpmqhw

Power Generation

As you already know, water is necessary for hydroelectric power generation. Hydropower plants harness the energy of flowing water to drive turbines that, in turn, generate electricity. It's one of the most reliable renewable energy sources, providing humans with power for decades.

Industrial Processes

Many modern-age industrial processes rely heavily on the use of water. It has numerous industrial applications, and it's widely used as a solvent in chemical reactions, for further processing of raw materials, in cooling systems that keep heavy machinery from overheating, and much more.

Construction and Building

Modern-day construction also relies heavily on water, essential in construction activities like mixing concrete, dust suppression, and settling foundations. It plays a critical role in creating infrastructure and buildings that form the backbone of communities.

Firefighting

Water is a primary tool in firefighting efforts. Fire hydrants, hoses, and relevant water sources are crucial for extinguishing fires and preventing their spread. Although fire-extinguishing substances like dry powders and carbon dioxide extinguishers are also used, various fire-extinguishing substances are water-based and use water in some form.

Understanding the diverse roles of water underscores its significance as a precious resource essential for sustaining life and supporting various aspects of human activities. However, with massive urbanization and irresponsible use of clean water, regions around the globe have started to face challenges regarding water supply. There's a growing need to update the current water supply and treatment systems to ensure the availability and conservation of water.

Water for Protection

In survival scenarios, where access to conventional hygiene practices may be limited, understanding how water is vital in personal cleanliness and wound care becomes imperative. Water becomes a crucial resource for first aid and emergency response. It's used for cleaning wounds, rehydrating individuals, and ensuring proper sanitation to prevent the spread of diseases. Here are some scenarios where the presence of water can make a difference.

Wound Cleaning

In medical emergencies and when giving first-aid, wound cleaning is the first step to preventing infection and promoting healing. Clean and uncontaminated water can effectively clean the site of injury, removing debris, harmful microorganisms, and dirt and priming the area for wound dressing. In medical terminology, rinsing the wound with clean or saline water is called wound irrigation, which reduces the risk of further infection and creates a suitable environment for the body's natural healing process.

Eye Irrigation

Like wound irrigation, eye irrigation is done when chemicals or foreign particles come into contact with the eyes. Water can flush out irritants and contaminants, providing rapid relief and preventing further damage. Continuous and gentle flushing removes every bit of the foreign substance. It minimizes the risk of injury as these particles can cause eye irritation, and when the eye is rubbed by force, it can damage the sclera and cornea, the outer layers of the eye.

Burn Treatment

Water is a fundamental component in the initial treatment of burns. Cool running water is applied to the burned area to help dissipate heat, reduce pain, and minimize tissue damage. Prolonged water application aids in cooling the burn and provides immediate relief. This immediate response with water is instrumental in the early stages of burn management when you are still waiting for professional medical attention. However, avoid using cold water or solid ice as it can exacerbate the burn site further.

Heat Exhaustion Mitigation

During hot days, heat exhaustion becomes a challenge that can be avoided by keeping yourself hydrated. In heat exhaustion, the body loses fluids, electrolytes, and minerals through sweating. This loss can be replenished by drinking water. However, in severe cases of heat exhaustion, people can become drowsy and even unconscious. In situations like these, alert the relevant authorities for medical assistance. While you wait for assistance, use a dampening cloth with water and gently rub it on the forehead, arms, legs, and stomach. This technique promotes evaporative cooling, lowering the body temperature and alleviating symptoms of heat-related illnesses.

Hypothermia Management

Water is integral in both preventing and managing hypothermia. Avoiding wet clothing and keeping the body dry in cold environments are essential to prevent heat loss. In cases where hypothermia has already set in, gradual warming with warm water to raise body temperature safely is the first step. Controlled water use is crucial to balancing the need for warmth without causing thermal shock to the body.

Decontamination from Chemical Exposure

Water is a primary component for decontamination in scenarios involving exposure to harmful chemicals. Rinsing the affected areas with copious amounts of water dilutes and removes the chemical substances, reducing the risk of further harm. Quick and thorough decontamination is critical to minimize the impact of chemical exposure and prevent the absorption of toxic substances into the body.

Oral Hydration in Shock

For individuals experiencing shock, maintaining proper hydration is crucial. In medical terms, shock can be triggered by severe trauma, allergic reaction, blood loss from an injury, or a heatstroke. When a person goes into a state of shock, the blood pressure drops drastically. Oral rehydration in this scenario stabilizes blood pressure and supports overall circulatory function, reducing the risk of organ damage. A sudden drop in blood pressure can also make the person unconscious. Avoid giving water if you notice the patient is feeling dizzy.

Splint Activation

In survival situations when you don't have immediate access to healthcare services, using emergency splints for a broken bone becomes crucial. These splints require water for activation and proper application. Emergency splinting is a temporary measure, but it's highly effective in keeping the injured bone stable until professional medical assistance is available. Water's role in activating bandages ensures proper adhesion and fitting, contributing to the stability of injured limbs.

Insect Bite Relief

It's a common first-aid practice for avid adventurers and explorers to use water after insect bites or stings. Running cold water on the area alleviates pain and reduces swelling. The cooling effect of water gives immediate relief and acts as a natural remedy in outdoor or wilderness first-aid scenarios.

Knowing the varied applications of water in first-aid scenarios emphasizes its critical role in providing immediate care, mitigating the impact of injuries, and contributing to the overall well-being of individuals in distress. Water is a versatile and essential resource in the first-aid toolkit, whether used for wound cleaning, temperature regulation, or chemical decontamination.

The Critical Role of Water

Besides assisting in first aid scenarios, there are various survival situations where water use becomes critical. Here are some of the diverse roles water is assigned, highlighting its necessity and the positive outcomes it delivers, even in the face of adversity.

Firefighting in Survival Situations

In survival scenarios, having the knowledge and ability to control and extinguish fires is crucial for safety and resource preservation. While using fire extinguishers is easier, it's best used for small fires. In case of a rapidly spreading fire, utilizing every method to minimize the spread of fire becomes necessary. Water, whether sourced from nearby rivers, lakes, or streams, can be carried in emergency supplies, becoming a primary tool for firefighting. It's used to douse flames, prevent the spread of fires, and protect essential resources like shelter and food supplies. In wilderness survival, where escape routes may be limited, effective firefighting can ensure your safety and the protection of your resources.

Signaling for Rescue

Water's reflective properties can be strategically utilized for signaling purposes in survival situations where you are stranded on or near a water body. Creating disturbances on the water's surface, such as waving or creating ripples, enhances visibility for potential rescuers. Mirrors or shiny surfaces can also be oriented to catch and redirect sunlight, creating visible signals. This method significantly increases the effectiveness of distress signals, making individuals more conspicuous to search and rescue teams or passing aircraft.

Water-Based Navigation Tools

Bodies of water serve as natural navigation guides in survival situations. Understanding water currents, watching the flow of rivers, and recognizing coastal features become valuable navigational tools. These water navigation techniques have been widely used for decades, assisting navigators to reach their destination without using modern geo-location gadgets like GPS. These water features offer insights into the lay of the land, aiding in decision-making for optimal navigation routes. This innate connection between water and navigation becomes a navigational advantage in diverse terrains.

Habitat Expansion

Although water's direct role in several situations is critical, groundwater and streams in an area promote habitat diversity. Water bodies attract various forms of wildlife. In survival scenarios, the presence of a water body points to the presence of nearby food sources. When water levels in these lakes and streams drop, it affects the groundwater tables, making them drop, limiting the water supply, and putting the habitat and nearby human population at risk. While water usage is a necessity for humans, making persistent conservation efforts ensures the streams, lakes, and rivers stay flowing.

Water's significance in survival tasks goes beyond its conventional roles. It becomes a dynamic and adaptable resource, offering solutions for firefighting, signaling, navigation, and resource procurement. Understanding the diverse applications of water in survival scenarios can empower you to use its potential creatively, contributing to your overall resilience and success in navigating and overcoming challenges in the wild.

Water in Crafting

Clay Molding and Construction

Pottery and using clay for construction are ancient techniques still being used today. Knowing how to mold clay and construct a shelter using essential Earth elements like water and clay can become a lifesaver in survival or emergencies. Water makes the clay malleable, so it can be shaped intricately. This simple yet effective technique can create containers, tools, and shelter.

Plant Processing and Fiber Extraction

Besides clay molding, water is crucial in processing plants for crafting materials. In survival crafts, plants are often used for weaving baskets, creating cordage, or constructing shelters. Water softens plant fibers, making them more pliable and easier to manipulate. Processed plant material is often used with clay to reinforce the construction and make it more resistant to weather changes. Furthermore, this soaking process enhances the plant fiber's flexibility, making it easier to weave into functional items like baskets and fiber-based containers. In survival situations, weaved baskets or containers can carry water, store food, or organize belongings.

Natural Dyeing and Coloring

Traditional crafts involve the art of natural dyeing, where water serves as the medium for extracting colors from plant materials. These decades-old dyeing and coloring methods can camouflage clothing, create signaling flags, or cater to specific functions in survival situations. The use of natural dyes, facilitated by water, adds a layer of resourcefulness to crafting.

Tanning and Leatherworking

Water makes the processing of raw hides into leather much easier. In the wild, this craft has been used by remote communities for ages, transforming raw hides into durable leather. In survival scenarios, this skill allows individuals to process animal hides for clothing, footwear, or improvised containers. Water softens hides, making them more pliable for crafting, and is integral to the tanning process, preserving the hides and transforming them into usable leather.

Cordage and Rope Making

Crafting cordage or rope from plant fibers is another valuable skill in survival scenarios. Water is used to soften and twist fibers, making creating strong and durable ropes easier. These ropes serve various purposes, from constructing shelters to crafting tools. Mastery of cordage-making enhances resourcefulness, allowing individuals to improvise essential items for different tasks.

Traditional Cooking and Food Preparation

Water is indispensable in traditional cooking as a critical component for boiling, steaming, and stewing. In survival situations, knowing how to purify water for consumption and use it for cooking safely is essential. Water ensures that safe and nourishing meals can be prepared and facilitates the extraction of nutrients from edible plants, making them more palatable and digestible.

Fire Craft and Fire-Hardening Tools

Water is strategically involved in firecraft and tool hardening. Traditional fire-starting methods often utilize materials that benefit from water treatment, making them more conducive to friction-based ignition. Likewise, tools crafted from wood or bone can be hardened through controlled exposure to water and fire, enhancing their durability and functionality in survival scenarios.

Natural Medicine and Herbal Infusions

Water is fundamental in traditional medicine and herbalism, serving as a medium for extracting beneficial compounds from plants. In survival situations, knowledge of medicinal plants and the ability to create herbal infusions or poultices using water becomes vital. Water helps extract therapeutic properties, allowing individuals to address minor ailments and injuries using natural remedies.

Environmental Awareness and Weather Prediction

Water's behavior in the environment holds valuable information for survival. To predict weather changes, traditional skills involve reading signs in water bodies, such as ripples, currents, or aquatic life. Understanding these cues aids in planning and adapting to environmental conditions, maximizing the chances of successful survival. It exemplifies how water is not just a resource but a source of information crucial for adapting to the ever-changing conditions of the outdoors.

Water's role in traditional skills and survival crafts goes beyond its physical properties. It catalyzes creativity and adaptability, allowing anyone to craft essential items, extract colors, and predict environmental changes. By understanding and utilizing water in these diverse skills, individuals enhance their resilience and resourcefulness in navigating the challenges of survival scenarios.

No doubt water is the elixir of life, and without it, life's existence becomes impossible. It's a one-of-a-kind substance, a driving force to living things, and a crucial element that is required in every aspect of life, from crafting to survival. The Earth is covered by 71% water, of which only 3% is fresh water. Out of this 3%, only 1.2% of water is drinkable. Although this precious liquid fueling life on Earth is still accessible, population expansion and urbanization threaten clean drinking water reserves on the planet. Fortunately, most governments worldwide are now taking effective steps to control water wastage and work on water conservation by planting more trees and creating urban green spaces. Many regions have even started monitoring water usage to gather relevant information and to take effective action to prevent the loss of this precious life source.

Bonus Chapter: Checklists

Now that you have worked through the book, this checklist will help guide your decisions on your water survival. The checklist covers all aspects of water use, including storage, purification, the equipment you need, and what your water will be used for. Use these guidelines as a quick reference to ensure you are on the right path. Sometimes, a lot of dense information can be overwhelming. This checklist can help you sift through all the data to ensure you have not missed any essential considerations.

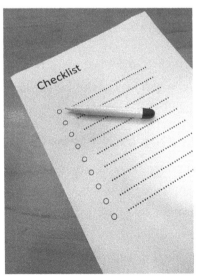

Tend to your water needs by keeping a checklist.
https://pixabay.com/photos/checklist-check-list-pen-3556832/

Whether you are setting up a homestead, traveling, prepping, or exploring the wilderness, an independent water source is needed. However, considering how dangerous it can be to consume and store water, you must double or triple-check that you are on point and have everything necessary for healthy, potable water. You do not want to find yourself in a position where you skim past an important detail that you already know. Even experts make mistakes, so having a comprehensive checklist to remind yourself of everything you must take into account is essential. Take your time and read through this checklist so that all your bases are covered and accounted for.

Hydro-Geography Checklist

- Have you considered the best way to collect water in your geographical location?
- What are the dangers affecting water in this area?
- How does the shape and structure of the land affect its water sources?
- Is the water in the area clean, or is it polluted?
- Is this region water scarce or abundant?
- Are there risks of flooding in the area?
- Are there seasonal droughts in the region?
- Does the water available in your area meet your daily requirements?
- What infrastructure or tools do you need to access the water?
- Is the water you can access on the surface or underground, and how will you gather it?

Rainwater and Dew Checklist

- How are you going to collect the rainwater?
- Which containers are best suited for your needs?
- What contaminants do you need to worry about in the area?
- Do you have a cool and dry place to store your water?
- Have you checked if your containers are leaking?
- Which water purification methods will you use to clean the rain or dew water?

- Is the surface you are using to collect rainwater clean?
- Will you boil your water or use other disinfection methods like bleach or chlorine tablets?
- Are you prepared for the dry season?
- Does your storage capabilities match your needs?
- Have you instituted cleaning protocols for yourself and the area where you store your water?
- Are your storage containers food-grade?
- Have you cleaned out your tanks?
- Do you rotate your water after six months?

Water Purification Checklist

- What kind of water filters are you using?
- Will you be making use of UV light technology to clean your water?
- Which kinds of chemical disinfection tablets meet your needs?
- Are your water purification methods portable so that you can carry them around?
- Have you considered how long you should boil your water by looking into the parasites and pathogens in your community or the region you are in?
- Have you checked for contaminants in your containers?
- Is your purification method suitable for what your water is being used for?
- Is your drinking water kept separated from water for other uses, and is it labeled?
- Do you have a water testing kit?

Snow and Ice Water Checklist

- Are you in an area where snow and ice are abundant?
- Is the ice made from freshwater or saltwater?
- Do you have enough fuel or energy to melt snow for your water consumption needs?
- What contaminants are in the snow, and how will you purify it?

Traveling with Water Checklist

- How much water do you need to carry, and do you have the capability to move that much?
- Are your containers durable?
- Do you have a way to boil the water?
- Have you got purification tablets?
- Where will you gather water when you travel, and is that source safe?
- Are your bottles or containers sealed properly?
- Have you got portable filtration devices?
- Do you have a UV purifier?
- Have you got an electric purifying device?
- Do you have a portable water testing kit?

Long-Term Water Storage Checklist

- Have your containers been used for anything other than water?
- Are your containers safe for water storage?
- Where will your water be stored, and what are you gathering water for?
- How long do you want to store water, and what material will the containers be made from that you will use?
- Do you have sufficient and appropriate space for the water?
- How many liters do you need to store?
- Does your water storage need to be mobile?
- How many people are making use of your water supply?
- Do you have the means to filter and disinfect your water?

Water Conservation Checklist

- Are your methods of water collection sustainable?
- Are the materials you are using environmentally friendly
- Have you been mindful of the chemicals you use and their effect on local ecology?
- Have you cleaned up after yourself?

- Have you responsibly used water without wasting it?
- Did you make considerations for the community around you when gathering water?
- Have you checked for leaks in your water storage and distribution systems?
- Did you consider your impact on the environment when you gather water?
- Are you using your fair share without overconsumption?

For all your water needs, work carefully through this checklist. Whether traveling with a small amount of water or storing hundreds of gallons over a long period, this checklist covers every aspect you should think about. The details of purification, storage, filtration, and distribution are too important to rush past. Take your time to work through this checklist and think deeply about what the prompts are pointing toward. Water can be the giver and taker of life, so it is important to respect this ancient substance. If you work well with water, it will meet many of your needs, but if you disrespect it, a disastrous outcome is inevitable.

Conclusion

Water is life. Humans, animals, birds, plants, and all other creatures need it to survive. If you plan to live off the grid, you must discover different water resources and learn how to collect, purify, and store the water for hydration, farming, bathing, etc.

The book starts by exploring the significance of water and its role in sustaining all life forms. You discovered how it is more vital than food and why it is always associated with life. You also learned the grave consequences of its deprivation.

You then understood the relationship between the physical characteristics of the land and the presence of water. You learned how to recognize mountain ranges, valleys, and other landforms. You uncovered how to locate water through its movement across landscapes.

You understood the natural process behind rain and dew formation. You then discovered the methodologies of collecting water through traditional and innovative techniques. Afterward, you learned to store the rain and dew water you collected. You learned about the best types of containers for long-term storage and their sizes and materials. You also saw the significance of keeping the water uncontaminated.

The collected water won't be safe for consumption until you purify it. You understood the risks of untreated water and learned different purifying methods so you and your family can consume clean water.

Most water resources will be frozen if you live in a cold environment. You discovered techniques for melting snow and ice. You also learned myths and misconceptions about snow, so you don't make any fatal mistakes.

If you live in an environment with limited water resources, you will need strategies to conserve water. You understood the challenges of a drought-prone region and learned effective techniques to minimize water usage.

People who are always on the move face more challenges than the ones who settle in one place. You learned about sourcing water when you are continuously mobile and your hydration needs due to increased physical exertion.

You then discovered techniques to keep your stored water supply uncontaminated and safe. You learned about the appropriate storage containers and their materials, designs, and capacities. You also discovered the best storage practices in various environments.

Water has many purposes besides hydration. You discovered the many uses of water in your everyday life. You also learned its significance in emergencies and survival tasks.

Life without water is impossible.

Here's another book by Dion Rosser that you might like

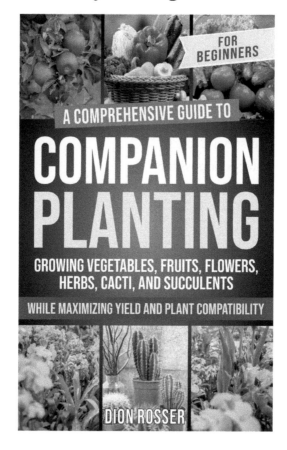

References

Revolutionizing urban spaces: 5 innovative rainwater harvesting techniques. (n.d.). Revolutionising Urban Spaces: 5 Innovative Rainwater Harvesting Techniques. https://smartwateronline.com/news/revolutionising-urban-spaces-5-innovative-rainwater-harvesting-techniques

(N.d.). Iwaponline.com. https://iwaponline.com/ws/article/20/8/3052/75992/Crafting-futures-together-scenarios-for-water

(N.d.). Masterclass.com. https://www.masterclass.com/articles/how-to-find-water

(N.d.-a). Amnh.org. https://www.amnh.org/explore/ology/water/what-is-water

(N.d.-a). Nyp.org. https://www.nyp.org/healthlibrary/definitions/untreated-water#:~:text=Untreated%20water%20is%20drinking%20water,and%20mild%20to%20severe%20illness.

(N.d.-b). Amnh.org. https://www.amnh.org/exhibitions/water-h2o--life/life-in-water/humans-and-water

(N.d.-b). Aspiringyouths.com. https://aspiringyouths.com/advantages-disadvantages/water-purifier/#google_vignette

(N.d.-c). Iwaponline.com. https://iwaponline.com/jwh/article/19/1/89/78374/Shungite-application-for-treatment-of-drinking

9 advantages of seawater desalination systems. (n.d.). Pure Aqua. Inc. https://pureaqua.com/blog/9-advantages-of-seawater-desalination-systems/

A solution to water scarcity. (n.d.). TREE AID. https://www.treeaid.org/blogs-updates/water/

Advantages & disadvantages of desalination. (n.d.). Brother Filtration. https://www.brotherfiltration.com/pros-and-cons-desalination/

Anderberg, J. (2016, April 20). How to find water in the wild. The Art of Manliness; Art of Manliness. https://www.artofmanliness.com/skills/outdoor-survival/how-to-find-water-in-the-wild/

Bramley, A. (2022, January 3). Is it safe to eat snow? Scientists say yes - with these caveats. NPR. https://www.npr.org/sections/thesalt/2016/01/23/463959512/so-you-want-to-eat-snow-is-it-safe-we-asked-scientists

Cal, R. (2019, March 16). 11 methods for off-grid water filtration and purification guide. Rustic Skills; Regina Cal. https://rusticskills.com/off-grid-water-systems/off-grid-water-filtration-purification/

Caldwell, J. (2019, July 11). Desalination Water Filtration Systems. What is Desalination? Water Equipment Technologies; Water Equipment Technologies - WET. https://wetpurewater.com/desalination-water-filtration-systems/

CDC. (2023, April 19). Creating and storing an emergency water supply. Centers for Disease Control and Prevention. https://www.cdc.gov/healthywater/emergency/creating-storing-emergency-water-supply.html

CDC. (2023a, April 13). Making water safe in an emergency. Centers for Disease Control and Prevention. https://www.cdc.gov/healthywater/emergency/making-water-safe.html

CDC. (2023a, April 13). Making water safe in an emergency. Centers for Disease Control and Prevention. https://www.cdc.gov/healthywater/emergency/making-water-safe.html

CDC. (2023b, April 13). Making water safe in an emergency. Centers for Disease Control and Prevention. https://www.cdc.gov/healthywater/emergency/making-water-safe.html

CDC. (2023b, April 19). Creating and storing an emergency water supply. Centers for Disease Control and Prevention. https://www.cdc.gov/healthywater/emergency/creating-storing-emergency-water-supply.html

Cho, R. (2011, March 7). The fog collectors: Harvesting water from thin air. State of the Planet; Columbia Climate School. https://news.climate.columbia.edu/2011/03/07/the-fog-collectors-harvesting-water-from-thin-air/

Clarke, J. (2022, December 10). Why can't you eat snow for hydration in a survival situation? advnture.com. https://www.advnture.com/features/dont-eat-snow

Complete guide to campervan water filtration. (2022, July 23). Engineers Who Van Life • DIY Van Building & Van Life.

https://engineerswhovanlife.com/campervan-water-filtration/

Cunningham, R. (2023, May 2). How to collect dew water: Efficient techniques for natural resource utilization. Survival World. https://www.survivalworld.com/water/gathering-dew/

Debutify, & Tasneem, S. (2022, January 3). 7 amazing shungite water benefits (2023 update). Atmosure. https://atmosure.com/blogs/stories/shungite-water-benefits

Dew. (n.d.). Nationalgeographic.org. https://education.nationalgeographic.org/resource/dew/

Drinking-water. (n.d.). Who.int. https://www.who.int/news-room/fact-sheets/detail/drinking-water

DrinkPrime. (2023, April 10). Boiled Water vs. Filtered Water: Which is Better? Drinkprime.In; DrinkPrime. https://drinkprime.in/blog/boiled-water-vs-filtered-water/

Fink, L. (2022, August 3). Clay water pot - the best way to filter your water in 2022. Uai Central. https://uaicentral.com/blogs/news/clay-water-pot

Fitzgerald, S. (2019, March 7). 6 plastic-free ways to travel with safe drinking water. National Geographic. https://www.nationalgeographic.com/travel/article/how-to-drink-water-safety-on-vacation-sustainability

Grooved surface accelerates dew harvesting –. (2019, March 20). Physics World. https://physicsworld.com/a/grooved-surface-accelerates-dew-harvesting/

Haas, E. (2018, September 18). The Hiker's Guide to Staying Hydrated and Treating Dehydration. Backpacker. https://www.backpacker.com/survival/how-to-stay-hydrated-and-treat-dehydration/

Hari, A. (2023, October 20). Can eating snow dehydrate you? Truth or myth. Medium. https://medium.com/@marketing_14327/can-eating-snow-dehydrates-you-truth-or-myth-d00b4fdb3882

Harvesting water and harnessing cooperation: Qanat systems in the Middle East and Asia. (n.d.). Middle East Institute. https://www.mei.edu/publications/harvesting-water-and-harnessing-cooperation-qanat-systems-middle-east-and-asia

Hitchcock, J. (2023, March 3). How to boil water without electricity - 15 easy ways. Survival Stoic. https://survivalstoic.com/how-to-boil-water-without-electricity/

Home water storage for an emergency. (2017, December 8). Utah Department of Environmental Quality. https://deq.utah.gov/drinking-water/emergency-water-storage

How to desalinate water. (2011, May 16). wikiHow. https://www.wikihow.com/Desalinate-Water

How to find water in a survival scenario. (2021, October 18). Operatorsassociation.com. https://www.operatorsassociation.com/how-to-find-water-in-a-survival-scenario

How to find water in a survival situation. (n.d.). Tacticalgear.com. https://tacticalgear.com/experts/how-to-find-water-in-a-survival-situation

HOW TO harvest rainwater. (n.d.). Off-Grid Collective. https://www.offgridcollective.co.nz/pages/how-to-harvest-rain-water

Hung, E. (2018, October 14). Best water storage containers for emergencies [tested]. Pewpewtactical.com; Pew Pew Media, Inc. https://www.pewpewtactical.com/best-water-storage-containers/

Impulse, S. (n.d.). Solutions to water scarcity. Solarimpulse.Com; Solar Impulse Foundation. https://solarimpulse.com/water-scarcity-solutions

Individual & Family Health. (n.d.). Bacteria, viruses, and parasites in drinking water. State.Mn.Us. https://www.health.state.mn.us/communities/environment/water/contaminants/bacteria.html

Isobeld. (2016, February 25). Melt water: How to get water from snow and ice. TGO Magazine. https://www.thegreatoutdoorsmag.com/skills/melt-water-how-to-get-water-from-ice-and-snow/

KPS. (2017, April 21). 10 ways to find water to survive the wilderness. Know Prepare Survive. https://knowpreparesurvive.com/survival/10-ways-to-find-water/

Kresh, M. (2018, October 14). Can clay jugs filter water? Green Prophet. https://www.greenprophet.com/2018/10/how-clay-jugs-make-polluted-water-safe/

Kylene. (2018, August 8). How to store water for emergency preparedness. The Provident Prepper - Common Sense Guide to Emergency Preparedness, Self-Reliance and Provident Living. https://theprovidentprepper.org/how-to-store-water-for-emergency-preparedness/

Lewicky, A. (2008). How to Melt Snow for water. SierraDescents. https://www.sierradescents.com/2008/05/how-to-melt-snow-for-water.html

Libretexts. (2020, May 27). 13.3: Water Scarcity and Solutions. Biology LibreTexts; Libretexts. https://bio.libretexts.org/Bookshelves/Ecology/Environmental_Science_(Ha_and_Schleiger)/04%3A_Humans_and_the_Environment/4.02%3A_Water_Resources/4.2.03%3A_Water_Scarcity_and_Solutions

List the uses of water in our daily life. (2022, July 4). Byjus.com; BYJU'S. https://byjus.com/question-answer/list-the-uses-of-water-in-our-daily-life/

Managing Water Scarcity. (n.d.). World Wildlife Fund. https://www.worldwildlife.org/projects/managing-water-scarcity

McKay, K. (2021, August 22). How to store water for long-term emergencies. The Art of Manliness; Art of Manliness. https://www.artofmanliness.com/skills/outdoor-survival/hydration-for-the-apocalypse-how-to-store-water-for-long-term-emergencies/

Miller, K. (2021, January 1). Is it safe to eat snow? Doctors explain possible side effects - prevention. Prevention.com. https://www.prevention.com/health/a34618470/is-it-safe-to-eat-snow/

Millhone, C. (2023, October 24). Shungite: Is this 'healing' stone as good for you as people say it is? Health. https://www.health.com/shungite-7972956

Minimising algal growth in farm dams. (2021, March 9). Agriculture Victoria. https://agriculture.vic.gov.au/farm-management/water/managing-dams/minimising-algal-growth-in-farm-dams

Mollah, M. (2023, May 23). 7 Sustainable Solutions to Water Scarcity —. Sea Going Green. https://www.seagoinggreen.org/blog/2023/5/23/7-sustainable-solutions-to-water-scarcity

Muskrat, J. (2017, October 16). Three ways to get safe drinking water from snow. Instructables. https://www.instructables.com/Three-Ways-to-Get-Safe-Drinking-Water-from-Snow/

Naude, J. (2022, November 1). How to Safely Store Water Long Term. Abeco Tanks. https://abecotanks.co.za/long-term-water-storage/

Offgrid Staff. (2017, July 26). The Myth of Water Rationing While Stranded in the Desert. RECOIL OFFGRID. https://www.offgridweb.com/preparation/the-myth-of-water-rationing-while-stranded-in-the-desert/

People, O., & Advisory Board. (2019, March 25). Understanding plastic recycling codes: Your guide to the RIC. Sustainable Brands. https://sustainablebrands.com/read/corporate-member-update/understanding-plastic-recycling-codes-your-guide-to-the-ric

Price, A. (n.d.). Obtaining water from snow and ice - dryad bushcraft. Dryadbushcraft.co.uk. https://www.dryadbushcraft.co.uk/bushcraft-how-to/obtaining-water-from-snow-and-ice

Rae. (2022, June 9). How to Make Fake Water for Crafts (+ Resin Terrarium Guide). Terrarium Tribe. https://terrariumtribe.com/fake-water-for-crafts/

Ramey, J. (2017, August 25). Best emergency water storage containers for your home. The Prepared. https://theprepared.com/homestead/reviews/best-two-week-emergency-water-storage-containers/

Ray, T. (2013, April 10). Water purification. American Hiking Society. https://americanhiking.org/resources/water-purification/

Reverse Osmosis. (2019, February 20). BYJUS; BYJU'S. https://byjus.com/chemistry/reverse-osmosis/

Reverse Osmosis. (2020, June 28). VEDANTU.
https://www.vedantu.com/chemistry/reverse-osmosis

Rosinger, A. Y. (n.d.). Human evolution led to an extreme thirst for water.
Scientific American.

Sahana. (2022, April 30). 10 advantages and disadvantages of chlorination of
water to know. Tech Quintal. https://www.techquintal.com/advantages-and-
disadvantages-of-chlorination-of-water/

Scavetta, A. (n.d.-a). Boiled water vs. Filtered water. Aquasana.com.
https://www.aquasana.com/info/boiled-water-vs-filtered-water-pd.html

Scavetta, A. (n.d.-b). Water Filter vs. Water Purifier: What is the Difference?
Aquasana.com. https://www.aquasana.com/info/water-filter-vs-water-purifier-
pd.html

Shock chlorination. (2020, August 5). Well Water Program.
https://wellwater.oregonstate.edu/well-water/bacteria/shock-chlorination

Shoop, M. (2010, September 10). How to filter water with clay pots. Sciencing;
Leaf Group. https://sciencing.com/filter-water-clay-pots-6975650.html

Singh, P. K. (2023, September 27). Water purification and its advantages and
disadvantages. Livpure. https://livpure.com/blogs/article/water-purification-and-
its-advantages-and-disadvantages

Sissons, C. (2020, May 27). What percentage of the human body is water?
Medicalnewstoday.com. https://www.medicalnewstoday.com/articles/what-
percentage-of-the-human-body-is-water

SITNFlash. (2019, September 26). Biological Roles of Water: Why is water
necessary for life? Science in the News.
https://sitn.hms.harvard.edu/uncategorized/2019/biological-roles-of-water-why-
is-water-necessary-for-life/

Smith, A. O. (2019, July 23). 10 effective ways to purify drinking water. A. O.
Smith India. https://www.aosmithindia.com/easy-and-effective-ways-to-purify-
water/

Solar water disinfection. (n.d.). Ctc-n.org. https://www.ctc-
n.org/technologies/solar-water-disinfection

Solar Water Disinfection. (n.d.). Thewatertreatmentplants.com.
http://www.thewatertreatmentplants.com/solar-water-disinfection.html

Solutions 12/01/2020, P. W. (2020, December 1). Is it safe to eat snow when
thirsty? Pentair Water Solutions. https://www.pentair.com/en-us/water-
softening-filtration/blog/snow-into-
water.html#:~:text=Collect%20ice%20or%20snow%20and,to%20catch%20the%
20falling%20water.

Store water safely - Hesperian health guides. (n.d.). Hesperian.org. https://en.hesperian.org/hhg/A_Community_Guide_to_Environmental_Health:Store_Water_Safely

Stricklin, T. (2023, October 5). 15 Dangerous Diseases Caused by contaminated drinking water. SpringWell Water Filtration Systems. https://www.springwellwater.com/15-dangerous-diseases-caused-by-contaminated-drinking-water/

Tallarico, G. (2018, July 20). Rainwater harvesting: 8 methods. World Permaculture Association. https://worldpermacultureassociation.com/rainwater-harvesting-8-methods/

The Best Ways to Stay Hydrated in the Wild. (n.d.). Survivor Filter. https://www.survivorfilter.com/blogs/home/the-best-ways-to-stay-hydrated-in-the-wild

TIMESOFINDIA.COM. (2023, August 25). Bringing back the Matka: Why clay pot water is the healthiest. Times Of India. https://timesofindia.indiatimes.com/life-style/food-news/bringing-back-the-matka-why-clay-pot-water-is-the-healthiest/articleshow/103057332.cms?from=mdr

Tom. (2023, February 2). Health risks of consuming melted snow water, what you should know. Weather Geeks.

Untreated water. (n.d.). Alberta.Ca. https://myhealth.alberta.ca/Health/Pages/conditions.aspx?hwid=stu3124&lang=en-ca

Us Epa, O. W. (2015). Emergency disinfection of drinking water. https://www.epa.gov/ground-water-and-drinking-water/emergency-disinfection-drinking-water

Us Epa, O. W. (2015). Emergency disinfection of drinking water. https://www.epa.gov/ground-water-and-drinking-water/emergency-disinfection-drinking-water

Us Epa, O. W. (2017). How we use water. https://www.epa.gov/watersense/how-we-use-water

UV Water Purification and How it Works. (n.d.). Espwaterproducts.com. https://www.espwaterproducts.com/understanding-uv-water-filtration-sterilization/

van Vuuren, A. (2023, June 28). The Advantages of Solar Powered Water Treatment Solutions. NuWater Water Treatment Solutions South Africa. https://nuwater.com/advantages-of-solar-powered-water-treatment-solutions/

Vuković, D. (2020, October 14). Food-safe plastics: Which plastic containers are safe for storing food and water? Primal Survivor.

https://www.primalsurvivor.net/food-safe-plastics/

Water conservation initiatives in drought-prone regions. (n.d.). Energy5.
https://energy5.com/water-conservation-initiatives-in-drought-prone-regions

Water in the wild. (2020, May 11). The Survival University.
https://thesurvivaluniversity.com/survival-tips/wilderness-survival-tips/all-things-water/water-in-the-wild/

Water purification and its advantages and disadvantages. (2018, April 23).
Jplast.Co.Za; JPlast. http://jplast.co.za/2018/04/23/water-purification-and-its-advantages-and-disadvantages/

Water Scarcity. (n.d.). World Wildlife Fund.
https://www.worldwildlife.org/threats/water-scarcity Lemire, M. (2023,
November 3). 5 Tips to Prevent Dehydration While Hiking. Adventure
Medical Kits. https://adventuremedicalkits.com/blogs/news/5-tips-to-prevent-dehydration-while-hiking

Water, C. (2022, December 13). What are the advantages and disadvantages of
different water filtration methods? Clean Tech Water.
https://www.cleantechwater.co.in/what-are-the-advantages-and-disadvantages-of-different-water-filtration-methods/

What are the benefits of a reverse Osmosis water filter? (n.d.).
Espwaterproducts.com. https://www.espwaterproducts.com/reverse-osmosis-advantages-and-disadvantages/

What is Hydrogeology, and what do Hydrogeologists do? (2018, December
17). IAH - The International Association of Hydrogeologists.
https://iah.org/education/general-public/what-is-hydrogeology

Why is water essential to life? (2020, January 9). Toppr Ask.
https://www.toppr.com/ask/question/why-is-water-essential-for-life/

Williams, T. (2023, June 27). Your Survival Guide on how to find Water in the
Wilderness. Desert Island Survival. https://www.desertislandsurvival.com/how-to-find-water/

Woodard, J. (2018, July 27). What is a reverse osmosis system and how does it
work? Fresh Water Systems.
https://www.freshwatersystems.com/blogs/blog/what-is-reverse-osmosis

Your guide to ion exchange water filters. (n.d.). Freedrinkingwater.com.
https://www.freedrinkingwater.com/water-education/quality-water-filtration-method-ion-exchange.htm

Zagala, R. (2018, March 20). Best off-grid water filtration: How to filter your
water without the power grid. The Berkey. https://theberkey.com/blogs/water-filter/best-off-grid-water-filtration-how-to-filter-your-water-without-the-power-grid

Zhou, W., Matsumoto, K., & Sawaki, M. (2023). Traditional domestic rainwater harvesting systems: classification, sustainability challenges, and future perspectives. Journal of Asian Architecture and Building Engineering, 22(2), 576–588. https://doi.org/10.1080/13467581.2022.2047979

Milton Keynes UK
Ingram Content Group UK Ltd.
UKHW022005280324
440346UK00005BA/54